Marine Paleobiodiversity

Marine Paleobiodiversity
Responses to Sea Level Cycles and Perturbations

Mu Ramkumar

ELSEVIER
AMSTERDAM • BOSTON • HEIDELBERG • LONDON
NEW YORK • OXFORD • PARIS • SAN DIEGO
SAN FRANCISCO • SINGAPORE • SYDNEY • TOKYO

Elsevier
Radarweg 29, PO Box 211, 1000 AE Amsterdam, Netherlands
The Boulevard, Langford Lane, Kidlington, Oxford OX5 1GB, UK
225 Wyman Street, Waltham, MA 02451, USA

ISBN: 978-0-12-805415-4

British Library Cataloguing-in-Publication Data
A catalogue record for this book is available from the British Library

Library of Congress Cataloging-in-Publication Data
A catalog record for this book is available from the Library of Congress

For information on all Elsevier publications visit
our website at http://store.elsevier.com/

Working together
to grow libraries in
developing countries

www.elsevier.com • www.bookaid.org

CONTENTS

PREFACE

Cognizant on the control exercised by the eustatic sea-level fluctuations over the stratal patterns, and the facies types and their distribution on spatial and temporal scales, the sequence stratigraphic concept was introduced in the 1970s. The studies that followed since then have established that the sedimentary facies successions were the results of earth processes that were on operation principally under the influence of global sea-level cycles. The concept also proposed that the processes create unique, hierarchically recognizable facies successions/bundles/parasequences and sequences. Though these sequences and sequence cycles of various orders and resultant changes in reservoir properties influenced by dynamics of depositional–erosional regimes, varied responses of siliciclastic, carbonate, and mixed siliciclastic systems to the relative sea-level fluctuations under the controls of tectono-eustasy, subsidence-uplift, sediment influx and climate, were studied extensively, the impacts and responses on ecosystem and biodiversity patterns, have largely been overlooked despite being in common knowledge for a long time.

Global sea-level changes were also related with paleoceanographic phenomena, such as oxygen depletion and water-mass stratification, which control biodiversity. Eustatic transgressions and regressions allowed connections/disconnections of ancient seas and oceans, and, thus, opened/closed migration routes and strengthened/diminished the occurrences and abundances of biota. Relative sea-level fluctuations, either eustatic or introduced by any other cause, either long term or short term or due to perturbations as a result of endogenic or exogenic processes, do exercise corresponding changes on the biotic system either directly by creating or depriving or expanding or contracting habitats on oceans as well as onland, and/or indirectly through atmospheric, lithospheric, and oceanographic processes due to their coupled nature. The results are explicit in terms of radiation and extinction events as well as in terms of abundance, dwindling, colonizing, acquiring new morphologic traits of species and habitat heterogeneity. Extensive review of the publicly available data and studies that documented various facets of paleobiodiversity trends covering Precambrian–Recent, from southern–northern latitudes, on a long-term, short-term, and perturbation scale perspectives, from arid–humid and dry–wet climes, in a wide ranging habitats spanning from highlands to bathyal regions, remains of ancient taxa as small as acritarch to as big as dinosauria and lifestyles ranging from sedentary, infaunal, epifaunal, nextonic, nektobenthic, avian, amphibious etc., and linking widely differing intrinsic, extrinsic, environmental, and other causes led to the proposition that, occurrence, abundance, and diversity of biota have responded dynamically to the prevalent changes in the earth's processes, despite the inherently episodic and discontinuous preservation.

Reviews on these aspects, as presented in this book, revealed that this episodic and discontinuous nature of biodiversity trends are akin to any other sedimentologic or geochemical proxies that are in use for inferring ancient environments. However, biotic proxies reveal more than other proxies could do; they reveal more on life: life that existed, that survived, and the conditions that led to survival and extinction! And, their presence and absence, both become evidences. Nothing more can be asked for and nothing of this sort can be expected from other proxies. Establishment of large databases at regional, global, faunal, and habitat specific natures, open access to them, the application of numeric methods on diversity data for a variety of scales and types, and the developments in removing biases have all provided unprecedented opportunity to document and analyze the spatio-temporal diversity trends engrained in the geologic past. Testing these concepts and methods on a variety of environmental settings recognized from strata of different ages and on different scales may provide clues for a better understanding, and thus, better future for this living fossil, the *Homo sapiens*.

Mu Ramkumar

ACKNOWLEDGMENTS

Understanding on this fascinating field of research was aided by the articles cited in this book that were the outcomes of many individuals and research teams who have toiled hard since many decades. Author places on record his deep sense of regard for all these authors. However, author alone is responsible for the views and statements presented. Interest to review and take stock of the current understanding on the paleobiodiversity and habitat dynamics as a result of sea-level fluctuations was driven by author's nomination to the National Working Group of IGCP-609 by the Geological Survey of India. Condice Janco, Publisher, Earth and Environmental Sciences, Elsevier Science and Technology Books; Louisa Hutchins, Acquisitions Editor; and other members of the editorial team of Elsevier, Prof. Vladan Radulović, Belgrade University and anonymous reviewers who have reviewed and recommended my proposal and those at the helm of administrative and technical approval committees at Elsevier are thanked profusely for their encouragement and continued support that helped hone my skills and scientific acumen. My involvement on this subject and scientific collaboration with national and international academic and research institutions has been and is being supported by research grants from various organizations, namely, the Alexander von Humboldt Foundation, Germany, the University Grants Commission, the Council of Scientific and Industrial Research, the Department of Science and Technology, the Oil Industry Development Board, the Oil and Natural Gas Corporation Limited, India, and the German Research Foundation, Germany, etc., for which I am thankful.

Many individuals, including, but not limited to, Prof. Vladan Radulovic, Faculty of Mining and Geology, Belgrade University; Prof. David Menier Jean Claude, University of South Brittany, France; Prof. Dmrity Ruban, Southern Federal University, Russia; Dr Jyotsana Rai, Birbal Sahni Institute of Palaeobotany, Lucknow, India; Dr Zsolt Berner, Institute of Mineralogy and Geochemistry,

Karlsruhe Institute of Technology, Germany; and Prof. Franz T. Fürsich, GeoZentrum Nordbayern, Fachgruppe PaläoUmwelt, Friedrich-Alexander-Universität Erlangen-Nürnberg, Germany, are thanked for encouragement and academic support. The team at Elsevier's production unit including Mohanapriyan Rajendran had done excellent work and is thanked for professional and flawless handling and transformation of draft manuscript into aesthetically designed and well-presented book.

My wife A. Shanthy, daughter Ra. Krushnakeerthana, and son Ra. Shreelakshminarasimhan are thanked for their support, understanding, and also for forbearing my absence while authoring this book. Above all, I submit my thankfulness unto the lotus feet of The Lord Shree Ranganayagi Samedha Shree Ranganatha, for his boundless mercy showered on me, and by whose ordinance those mentioned above have shouldered the responsibilities either actively or passively and helped me to complete this work.

Mu Ramkumar

CHAPTER 1

Introduction

Cognizant on the control exercised by the eustatic sea-level fluctuations over the stratal patterns and the facies distribution, the sequence stratigraphic concept was introduced. It provided an impetus for predicting the facies pattern in spatial and in part at temporal domains and correlating the stratigraphic records with their counterparts located elsewhere. The relative sea-level cycles, first published by Vail et al. (1977) and revised by Haq et al. (1987, 1988) and more recently (in part) by Haq (2014) espoused that the sedimentary sequences are produced principally under the influence of eustatic sea-level cycles of few tens of millions (first-order cycle) to few million (third-order cycle) year durations. The relative sea-level chart based on the seismic stratigraphic studies at Exxon, first presented by Vail et al. (1977) and later revised by Haq et al. (1987, 1988) recognized about 100 global sea-level changes. Mitchum Jr. and Van Wagoner (1991) reported that the interpreted eustatic cyclicity has a pattern of superposed cycles with frequencies in the ranges of 9–10, 1–2, 0.1–0.2, and 0.01–0.02 Ma (second- through fifth-order cyclicity, respectively). These sequence cycles showed a hierarchical stacking pattern such that fifth-order shallowing upward cycles grouped into fourth-order cycles, which in turn stacked vertically into part of a third-order cycle. This pattern was recognized by many authors in sedimentary facies successions (Goldhammer et al., 1991; McGhee Jr, 1992; Gjelberg and Steel, 1995) as well as independent geochemical analyses (Veizer, 1985; Veizer et al., 1997, 1999, 2000), lithofacies-based paleobathymetric estimates (Ramkumar et al., 2004), and integrated multiproxy estimates (Ramkumar et al., 2011; Ramkumar, 2015).

Hays et al. (1976) convincingly demonstrated that the climatic records were dominated by frequencies characteristic of variations in the Earth's tilt, precession and eccentricity relative to the Sun. Successive studies have shown that distinct sedimentary sequences

Marine Paleobiodiversity. DOI: http://dx.doi.org/10.1016/B978-0-12-805415-4.00001-2

could be traced to sea-level cycles of up to infra seventh order (Nelson et al., 1985; Williams et al., 1988; Carter et al., 1991; Tucker et al., 2009). In the years since, numerous studies have upheld the notion of occurrences of cyclic sedimentation under the influence of the short-term Milankovitch glacio-eustatic and climatic cycles in terms of 100, 41, and 23 ka orbital periods (Laferriere et al., 1987; Grammer et al., 1996; Gale et al., 2002) that influence or control global ice volume, thermohaline circulation, continental aridity and run off, sea surface temperature, deep ocean carbonate preservation, and atmospheric CO_2 and methane concentrations (Raymo et al., 1997). These studies have also established that the earth processes were on operation at various timescales to create unique hierarchical sedimentary facies succession principally under the influence of global sea-level cycles.

In addition to changing facies distribution and stacking pattern, global sea-level changes might also altered the paleoceanographic phenomena, such as oxygen depletion and water-mass stratification, which were important controls on biodiversity. The eustatic transgressions and regressions might have allowed connections/disconnections of ancient seas and oceans, and, thus, opened/closed migration routes and strengthened/diminished the occurrence and abundance of biota. As Ruban (2010a) has demonstrated, the eustatic fluctuations were responsible for the dynamics of ecological niches. The present environments and ecosystems are an evolving continuum with those of the past and the future. The term "ecosystem" denotes the biological community together with the abiotic environment in which it is set as both are strongly linked to each other by fluxes of energy and matter (Matsukawa et al., 2006). Hence, in order to understand this continuum, estimations on geological–biological dynamics through biotic abundance (Fioroni et al., 2015) are essential. Coordinated study on the responses of biota to environmental parameters associated with global change has ecological and geological implications (Hallock, 2005).

Thus, this book is an attempt to introduce the readers the current trends of biodiversity studies with an emphasis on the importance of documentation of biodiversity and habitat response to eustatic and other sea-level fluctuations. Though the sea-level fluctuations and resultant changes in environmental settings and by implication, attendant changes in spatio-temporal variations of ecological niches, habitats and population and diversity of fauna and flora that were dependent/influenced by specific set of environmental conditions were known for a long time, appreciation of these phenomena, and attempts to understand them in geological perspective have only commenced very recently. Thus, this book is intended for beginners, advanced level undergraduate, postgraduate and those interested in understanding the biotic responses to sea-level fluctuations. However, it is also brought to the attention that this book is not the inventory of previous studies on these aspects; instead, only those articles that most closely served the purpose of subject matter are cited. For further reading and for more advanced treatment, readers are suggested to consult the articles cited and the references therein.

CHAPTER 2

Rationale

The study of Tibert and Leckie (2004) is a case in point where they have demonstrated the distinct vertical partitioning of taxon associations with distinct changes between high-tide and low marsh-tide, i.e., at an altitudinal change of 1 m. It also emphasizes the magnitude of impact the sea-level fluctuations (however minor may be) could cause on the abundance and diversity of biota. The microfauna, especially foraminifera, like many other benthic inhabitants of marine habitats, are highly sensitive to several physical, chemical, and several other oceanographic parameters such as nature and stability of substrate, bathymetry, energy, salinity, oxygen content, and dissolved organic carbon content. These traits are successfully exploited by integrated analysis of sediment properties and faunal content, to accurately estimate the paleobathymetric oscillations. There are few large databases established for this purpose. Rossi and Horton (2009) have evaluated the efficiency and accuracy of one such database to find that it is possible to reconstruct paleobathymetric variations based on integrated sedimentological and foraminiferal data, and the predictability of the database is more accurate ($R^2 = 0.95$) in depth range between 8 and 100 m.

Though the sequence cycles of various orders and resultant changes in reservoir properties influenced by dynamics of depositional–erosional regimes, and the varied responses of siliciclastic, carbonate, and mixed siliciclastic systems to the relative sea-level fluctuations under the varied, yet relative influences of tectono-eustasy, subsidence-uplift, sediment influx and climate were documented, the impacts and responses of ecosystem and biodiversity patterns, despite being in common knowledge for a long time (Newell, 1967; Valentine, 1968; Hallam, 1977), have largely been overlooked except few excellent studies (Hallam and Wignall, 1999; Gale et al., 2000; O'Dogherty et al., 2000; Sandoval et al., 2001a,b; Yacobucci, 2005; Purdy, 2008; Ruban, 2010a,b; Melott and Bambach, 2011; Holland, 2012; Masse and

Marine Paleobiodiversity. DOI: http://dx.doi.org/10.1016/B978-0-12-805415-4.00002-4

Fenerci-Masse, 2011, 2013). However, these publications are minuscule when compared with the myriad varieties of habitats prevalent during the geologic past, innumerable cycles and perturbations of sea levels, and countless biota and their community structures. These studies are only few examples among the innumerable sequence and implied reservoir studies where the occurrences of sea-level cycles are documented, but the associated biodiversity and ecosystem dynamics are overlooked. Thus, through extensive review of the published literature, this book attempts to

a. elucidate the current trends in paleobiodiversity studies,
b. examine the causes and parameters that are related to the observed biodiversity trends,
c. evaluate the long term, short term, and perturbations in biodiversity and causative mechanisms,
d. understand the relative influence of relative sea-level fluctuations on biodiversity and habitat heterogeneity,
e. suggest future course of action in the light of inferences that emanate from the review.

CHAPTER 3

Geological Perspective on Biodiversity and Habitat

3.1 TERMINOLOGIES

As the term *diversity* is often used to indicate the number of taxa within a given sample and to a metric incorporating both the number of taxa and its relative abundance within a sample, Sessa et al. (2012) recommended to use the term *richness* to refer the number of taxa, *evenness* to reflect relative abundance, and *diversity* to signify both. At the interspecies level, *structural biodiversity* can be estimated by *specific richness* and by *morphological diversity* (also called *disparity*). Disparity depicts morphological changes and assesses the underlying processes and is a relevant proxy to account for diversity changes through time in terms of taxonomy, population differentiation, developmental constraints, and ecological stresses (Laffont et al., 2011). Prior to any disparity study, organism forms must be quantified, for which, *morphometrics* is employed. It deals with shape analysis and shape variation among specimens of a single *population* or among specimens of distinct populations (Laffont et al., 2011).

3.2 TYPES AND COMPONENTS

Sepkoski Jr. (1988) defined biodiversity into three components, namely, α, β, and γ that describe intracommunity (closer packing of species within existing communities), intercommunity (occupation of new ecospace), and interprovince (dispersal of the continents and terranes). According to Laffont et al. (2011), biodiversity can be described at three levels, namely, at the population, the interspecies, and the ecosystem levels that account for diversity of genes within a species, among the species, and at assemblage/community levels, respectively.

Marine Paleobiodiversity. DOI: http://dx.doi.org/10.1016/B978-0-12-805415-4.00003-6

3.3 GEOLOGICAL PERSPECTIVE

Reconstructing diversity dynamics prevalent during the geologic past is the essential task of paleobiology (Svenning et al., 2011; Newham et al., 2014). There is a long history of diversity counts of fossil organisms in order to understand the evolution of life through geological time (see Servais et al., 2010, for a review on the history of paleobiodiversity studies). Understanding how taxonomical biodiversity originates and maintains over a wide range of time and spatial scales has become a major objective in ecological and evolution research due to major concerns over the future of the Earth's biodiversity. The paleontological record offers a tremendous amount of information for addressing these questions over large time and spatial scales, including the interplay between biotic (i.e., ecological and evolutionary) and abiotic processes such as environmental conditions and climatic fluctuations and their effect on past and present biodiversity (Perrin and Bosellini, 2012), which in turn have significant relationships and implications on the studies of sedimentary basins in the context of sequence formation. Though it is of common knowledge that the regional transgressions and regressions influenced marine biodiversity, the extents, differences among various biota and ecological niches and also during different time slices are still poorly understood (Ruban, 2007).

3.4 THE LINK BETWEEN GEOLOGICAL PROCESSES AND BIODIVERSITY

Changes in sea level would change the area of epicontinental seas, which directly determines area available for both biotic habitats and sediment accommodation in similar ways (Newell, 1952; Simberloff, 1974; Crampton et al., 2011). Peters (2008) demonstrated the relationship between Phanerozoic marine biodiversity and the temporal and spatial distributions of carbonate and siliciclastic lithofacies. A link between habitat heterogeneity and taxonomic diversity is another way in which the geologic processes might directly influence living organisms to yield correlated patterns in the rock and fossil records. Habitat heterogeneity over geographic space is a strong determinant of taxonomic richness (Rook et al., 2013).

Retallack (2011) and Sigwart et al. (2014) listed rapid burial (obrution), stagnation (eutrophic anoxia), fecal pollution (septic anoxia), bacterial sealing (microbial death masks), brine pickling (salinization), mineral infiltration (permineralization and nodule formation by authigenic cementation), incomplete combustion (charcoalification), desiccation (mummification), and freezing as the criteria/events that supported exceptional preservation of paleotaxa in abundance. Based on worldwide search, Retallack (2011) also documented that times of especially widespread exceptional fossil preservation were also times of stage boundaries, mass extinctions, oceanic anoxic events, carbon isotope anomalies, spikes of high atmospheric CO_2, and transient warm-wet paleoclimates in arid lands. Cherns et al. (2008) opined that if the initial microstructural and mineralogical characteristics of certain organisms such as molluscs that preferentially get destroyed (Plate 3.1a,b) and their fossilization potential is less than many of the contemporary organisms (Plate 3.1c,d) and skew the biodiversity computations, the use of storm beds (Plate 3.2a,b), shell plasters and submarine hardgrounds (Plate 3.2c,d) and low energy, organic-rich mud-dominated settings, can be identified as fossil deposits that can

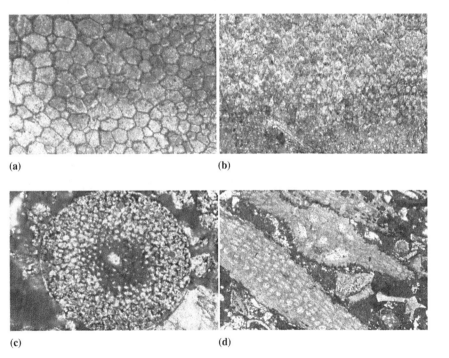

(a)   (b)

(c)   (d)

Plate 3.1 (a) Photomicrograph showing well-preserved prismatic internal structure of molluscan shell. (b) Photomicrograph showing neomorphically altered prismatic structure of molluscan shell. (c) Photomicrograph showing neomorphically altered internal structure of echinoderm plate. (d) When compared with other organisms, the foraminiferal tests and their shell wall internal structures are relatively well preserved as depicted in the photomicrograph.

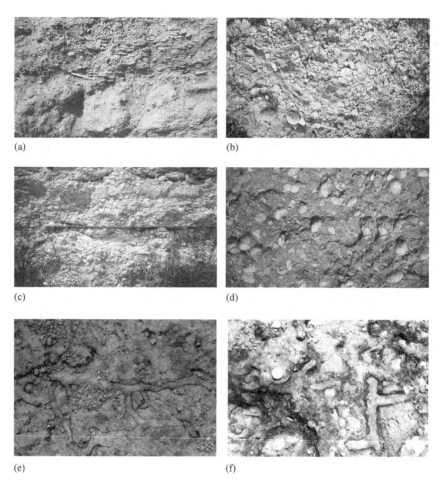

Plate 3.2 (a) Field photograph showing the occurrence of paleostorm deposits that contain a highly diverse macro and microfossil community in a relatively less thick bed. Recognition of paleostorm deposits (Ramkumar, 2006) and the biotic content contained in them offer unique taphonomic windows. Location of the photograph: TANCEM quarry I, near Kallankurichchi village in the Ariyalur area of the Cauvery Basin, South India. (b) Planar view of the bed described in the previous photograph showing the thick population of fragmented diverse shell remains as well as whole tests of irregular echinoderm Stigatophygus elatus. *(c) Field photograph showing the occurrences of hard ground surfaces (Ramkumar, 1996), associated to which, thick population of biota occur. Location of the Photograph: TANCEM Mines III, near Kallankurichchi Village, Ariyalur area, Cauvery Basin, South India. (d) Planar view of the exposure described in the previous photograph. Note the thick population of* gryphea. *(e) Field photograph showing the burrow systems of Ophiomorpha irregulaire (Ramkumar and Sathish, 2009) with branching nature. The horizontal tunnels are more or less uniform in diameter and show branching at right angles. The causative organisms of these burrows, possibly decapods colonized the ecospaces vacated by the previous occupants soon after prevalent storm. This inference shows that, with major environmental shift and attendant environmental changes, the biotic composition also changes significantly. Location of the photograph: TANCEM mines II, northern minefloor, near Kallankurichchi Village, Ariyalur area, Cauvery Basin, South India. (f) Closeup view of the previous photograph. Field photograph showing right angle branching of tunnels of Ophiomorpha irregulaire. The photograph also shows the differential nature of burrow fill and the burrow lining as explicit in terms of differential tones.*

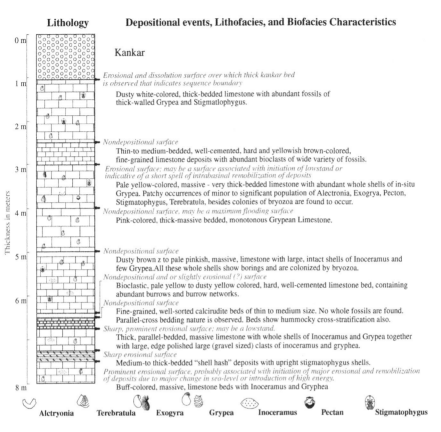

Lithology   Depositional events, Lithofacies, and Biofacies Characteristics

Kankar

Erosional and dissolution surface over which thick kankar bed is observed that indicates sequence boundary
Dusty white-colored, thick-bedded limestone with abundant fossils of thick-walled Grypea and Stigmatlophygus.

Nondepositional surface
Thin-to medium-bedded, well-cemented, hard and yellowish brown-colored, fine-grained limestone deposits with abundant bioclasts of wide variety of fossils.
Erosional surface; may be a surface associated with initiation of lowstand or indicative of a short spell of intrabasinal remobilization of deposits
Pale yellow-colored, massive - very thick-bedded limestone with abundant whole shells of in-situ Grypea. Patchy occurrences of minor to significant population of Alectronia, Exogrya, Pecton, Stigmatophygus, Terebratula, besides colonies of bryozoa are found to occur.
Nondepositional surface, may be a maximum flooding surface
Pink-colored, thick-massive bedded, monotonous Grypean Limestone.

Nondepositional surface
Dusty brown z to pale pinkish, massive, limestone with large, intact shells of Inoceramus and few Grypea.All these whole shells show borings and are colonized by bryozoa.
Nondepositional and or slightly erosional (?) surface
Bioclastic, pale yellow to dusty yellow colored, hard, well-cemented limestone bed, containing abundant burrows and burrow networks.
Nondepositional surface
Fine-grained, well-sorted calcirudite beds of thin to medium size. No whole fossils are found. Parallel-cross bedding nature is observed. Beds show hummocky cross-stratification also.
Sharp, prominent erosional surface; may be a lowstand.
Thick, parallel-bedded, massive limestone with whole shells of Inoceramus and Grypea together with large, edge polished large (gravel sized) clasts of inoceramus and gryphea.
Sharp erosional surface
Medium-to thick-bedded "shell hash" deposits with upright stigmatophygus shells.
Prominent erosional surface, probably associated with initiation of major erosional and remobilization of deposits due to major change in sea-level or introduction of high energy.
Buff-colored, massive, limestone beds with Inoceramus and Gryphea

Alctryonia   Terebratula   Exogyra   Grypea   Inoceramus   Pectan   Stigmatophygus

Figure 3.1 Depositional events of the Maastrichtian section, TANCEM mine, Ariyalur area, Cauvery Basin. Temporal variations of Lithofacies, biofacies, and faunistic compositions, due to major storm during Maastrichtian in the Cauvery Basin (after Ramkumar, 2006 and Ramkumar and Sathish, 2009), are depicted and described in the figure. This depiction also demonstrates that simple, systematic documentation of litho and biofacies characteristics, internal sedimentary structures, contact relationships between facies types and the faunistic composition contained in them could reveal the biodiversity and habitat heterogeneity.

preserve the labile aragonitic component of the fauna and thus these may represent potential taphonomic windows. Major and sudden environmental changes also destroy ecological niches favorable for certain organisms that bring in opportunistic colonizers (Plate 3.2e,f). As could be appreciated from Figure 3.1, the changes in prevalent environmental conditions always impact the biotic composition.

Opportunistic colonization as a result of changes in environmental conditions associated with sea-level fluctuations, not necessarily associated with that of catastrophic events, has been excellently documented by Reolid et al. (2015), while proliferation and abundance of selected nannofossil taxa associated with major paleoclimatic changes were reported by Fioroni et al. (2015). While examining the trends of nannotaxa abundance, geochemical and climatic data, in relation with the fluctuations of carbonate compensation depth which in turn was under the influence of icehouse–greenhouse reversals, Fioroni et al. (2015) observed a strong relationship between nannofossil abundance and the climatic switches. Integrated analysis of foraminiferal assemblages, geochemical proxies, in the Oued Bahloul section (Tunisia) allowed these authors to reconstruct the environmental turnover across the Cenomanian–Turonian boundary. These authors also documented prevalent eutrophic conditions evidenced by high proportions of buliminids and the replacement of planktic oligotrophic specialist *Rotalipora* by eutrophic opportunist *Hedbergella*.

There are certain unique traits of certain organisms and fossilization events that promote exceptional preservation, despite intrinsic microstructural and mineralogic traits of shells, such as those of ammonites. Though composed of predominantly aragonite and high magnesian calcitic mineralogy and thin fibrous/prismatic microstructure, and have less stability than low magnesian calcitic brachiopods, the ammonites display exceptional properties such as high abundance, widespread occurrence, high evolutionary rates, high taxonomic diversity, and morphological disparity (De Baets et al., 2012, 2013). Other conditions being equal, such selective representation of few organisms than other contemporaneous organisms may influence the biodiversity patterns. Owing to their occurrences in well-known stratigraphic framework (e.g., ammonites, Foote and Sepkoski, 1999), the bias may get further exacerbated unless proper understanding and holistic background patterns are taken into consideration during biodiversity analysis. In one of the most detailed and critical examination of pristine nature and preservation characteristics of fossil shells, Ullmann et al. (2015) documented the microstructure, cathodoluminescence images, and elemental and stable isotopic composition individual growth bands of belemnite rostrum (Plate 3.3a) and presented a protocol for using the belemnite rostrum based chemostratigraphic and paleoclimatic interpretations. The implications of this study for paleobiodiversity/ fossilization potential of selective biota is that, in addition to the lifestyle of the organisms, nature of their habitat and the depositional and diagenetic conditions that took place prior to, during and after the death of the organisms, other geological processes are also biased

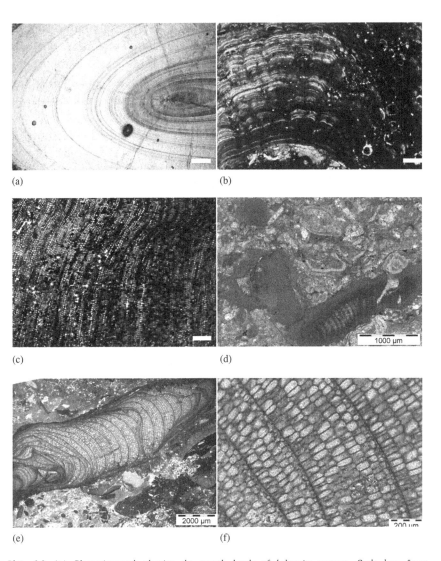

Plate 3.3 (a) Photomicrograph showing the growth bands of belemnite rostrum. Scale bar: 5 mm. (b) Photomicrograph showing the microstructure of stromatolite growth bands. Scale bar: 5 mm. (c) Photomicrograph showing the microstructure of petrified wood. Scale bar: 5 mm. (d) Photomicrograph showing the preservation of rhodolith and preferentially recrystallized other shell fragments in the vicinity. (e) Photomicrograph showing the excellently preserved microstructure and growth bands of coralline algae. (f) Closeup view of the specimen shown in previous photograph, depicting the microstructure of coralline algae.

against exceptional preservation and only in extraordinary circumstances, true signals of past biodiversity could be revealed. The study of Halfar et al. (2000) is an example, where certain biota such as rhodolith (Plate 3.3d–f), despite inheriting geologically unstable

microstructural and physical traits and inhabiting dynamic depositional settings (Littoral-Neritic zones of tropic-polar regions), are well
represented in the fossil record than other biota (e.g., compare the
photomicrographs in Plate 3.3), and found suitable for interpreting
ultrahigh resolution paleoceanographic and climatic conditions. See
Jakubowicz et al. (2015) for a detailed description of protocols and
requirements involved in selection and use of fossil skeletons to
interpret paleoecology.

CHAPTER 4

Methods of Biodiversity and Habitat Heterogeneity Analyses

The nature of the stratigraphic record and limited access to the parameters defining fossil organisms put specific constraints on the study of past taxonomic diversity. Species have responded to past environmental changes in individualistic and complex manners, in terms of tolerance, evolutionary adaption, migration, and local and global extinctions. Modeling past dynamics of species distributions is therefore not a simple endeavor, but requires careful integration of data and methods as well as considered handling of assumptions and limitations (Svenning et al., 2011). In this chapter, methods used in paleobiodiversity analyses, their limitations, and the techniques employed overcome the shortcomings are presented.

4.1 APPROACHES OF BIODIVERSITY

Fara (2004) stated that estimates of past diversity using paleontological evidence can be achieved within two main analytical frameworks. The traditional taxic approach (*sensu* Levinton, 1988) makes a straightforward use of stratigraphic ranges at a given taxonomic level, whereas the phylogenetic approach (*sensu* Smith, 1994) derives paleodiversity estimates from both phylogenetic topologies and stratigraphic occurrences (see Fara, 2004, for a detailed discussion on these two approaches). Diversity counts can be achieved by a simple, intuitive protocol that incorporates all fossil occurrences and minimal principles of evolutionary continuity. When empirical occurrence data are artificially degraded, the technique captures an equal or a better part of the original diversity signal than the taxic approach.

Marine Paleobiodiversity. DOI: http://dx.doi.org/10.1016/B978-0-12-805415-4.00004-8

4.2 NUMERICAL METHODS AND DATABASE-BASED ANALYSES

Monnet (2009) used several metrics to extract and analyze the biodiversity patterns. The *species richness* is defined as the number of species occurring within a given zone. *Origination* and *extinction* values correspond to the number of taxa appearing (First Occurrence, FO) and disappearing (Last Occurrence, LO) between two successive association zones. The percentage of origination is defined as the number of FOs divided by the total number of taxa occurring in the next overlying zone. The *turnover* is defined as the sum of the number of originations and extinctions. The percentage of turnover corresponds to the turnover divided by the total number of distinct taxa present in the two bracketing zones. Depending on the requirement, availability of requisite data and other limitations, either distinction is made between pseudo (FO, LO) and true originations/extinctions (First Appearance Datum, FAD; Last Appearance Datum, LAD) or the *trends* are analyzed without making any distinction.

Though biodiversity trends were analyzed during 1860s based on the Darwin's treatise (Servais et al., 2010), application of numerical methods for understanding taxonomic diversity commenced only in the 1970s (see Cascales-Miñana et al., 2013, for a review). One of the basic methods to analyze diversity is to plot the number of known/reported/taxonomically described species/genus/taxon against chronological datum and interpret the trends. For example, Tappan and Loeblich Jr (1973) plotted the number of described species of acritarchs per geological period, showing their predominance among major phytoplanktonic groups during Paleozoic times, their increasing diversity from Precambrian to Silurian followed by a marked decrease at the Devonian–Carboniferous boundary, and their replacement by the dinocysts from the Late Triassic to the present. Similarly, Strother (1996) plotted the number of all validly published genera per Million annum (Ma). This distribution of genera over time confirmed the general temporal trend already described by Tappan and Loeblich Jr (1973). According to Ruban (2010b), biodiversity curves have to be inclusive of the absolute time scale such as the use of modern chronostratigraphic framework without important errors (ie, the expected maximum error should not exceed the duration of trends and events recognized in the study). Major *radiations* are interpreted when an acceleration of the taxa number by no less than 1.5–2.0 times during a single episode delineated by a clear trend with a more or less constant speed of change (ie, constant angle of biodiversity curve) and no significant interruptions as significant increases in the marine biodiversity. When possible, other geological/paleontological data have to be included in a supportive role to justify the event, the timing, and the quantum of radiation/extinction. The efficiency of different metrics and statistical methods that attempt to characterize biodiversity was analyzed by Villier and Navarro (2004) through multivariate statistical methods.

Vecoli and Hérissé (2004) provided the methods of computation of different *diversity metrics*. According to them, diversity metrics can be calculated for a series of 1 or 2 million year-long time slices into which the entire study period should be subdivided to reduce errors due to unequal duration of stages (for stage-level diversity calculations) or biozones (for biozone-level diversity). This assumes a restrictive linear relationship, which is not necessarily correct and does not remove all undesirable statistical properties of diversity estimates and hence regressions have to be used to standardize data. The analysis of residuals allows a corrected and detrended signal to be constructed. The accuracy of standardized data depends on the statistical robustness of the regression model (Villier and Navarro, 2004). *Total diversity* is calculated as the total number of taxa (genera and species) co-occurring during each chronostratigraphic interval. *Continuing diversity* considers the number of species crossing one interval to the next. In this case, the change between the number of species entering an interval and leaving that interval can provide a good evaluation of the change in diversity during that time independently from preservation or spatial biasing effects. *Modified diversity* involves counting the number of species ranging through the time unit (from the preceding to the following) plus half the number of species ranging beyond the time unit but originating or ending within it, plus half the number of species confined to the time unit. It is also common to calculate and plot origination and extinction rates simply as the number of species originating or going extinct during a time interval, divided by the duration of the interval (1 or 2 million years).

Kiessling (2002) examined the global diversity patterns through the use of different indices such as the *Margalev's index*: $DM = (S - 1)/\ln(N)$, where S is the number of species and N is the number of individuals in a sample or region. This index provided a measure of species richness that is roughly normalized for sample size without using more complex *rarefaction* techniques. Rarefaction effect is the bias that may result due to the artifact in proportion to the sample size. Use of the formula provided by Hurlbert (1971) is suggested by Aguirre et al. (2000) to remove this bias before attempting any fruitful diversity analysis.

Commonly used rarefaction techniques include, but not limited to: the *Shannon index*: $H' = -\Sigma pi X 1n(pi)$, where pi is the fraction of the ith species of the total fauna. This index provide a rough measure of diversity, which is also less biased by sample size than species richness and consider relative abundance data; the *Evenness*: $J = H'/H'_{max}$ where H' is the Shannon index as defined above and $H'_{max} = \ln S$. This index determines how evenly the proportions of taxa are distributed in a sample and thus independent of sample size and species richness; the *Simpson's Index of Diversity*: $1 - (\Sigma n(n-1))/(N(N-1))$, where n is the number of specimens of a species and N is the total number of specimens (Simpson, 1949). This measure accounts not only for the number of specimens involved but also for the number of specimens per species (Klompmaker et al., 2013). This index ranges close to 1 when the species in a community are distributed homogeneously and drops towards 0 when one or few species dominate the community (Perrier et al., 2012).

To calculate the theoretical number of species if an infinite number of specimens had been collected, the *Chao1 Index* (Chao, 1984) $S_{Chao1} = S_{obs} + (F_1^2/2 \ F_2)$ is utilized, where S_{obs} is the number of species in the sample, F_1 is the observed number of species represented by one specimen, and F_2 is the observed number of species represented by two specimens. It estimates the absolute number of species in an assemblage (Klompmaker et al., 2013). There are many other metrics, such as the Simpson Coefficient, the Jaccard Coefficient, and the Dice Coefficient or the Sørensen similarity index, for analyzing the number of species in each sample to treat all weightages of all the species equally while the metrics, such as the Morisita–Horn Index, the Relative Abundance Index, the Bray–Curtis Dissimilarity Index, and the Yue and Clayton theta similarity coefficient, take into account the relative abundance of each species (Klompmaker et al., 2013).

Initially, paleobiologists followed a simple procedure of counting the known higher taxa in each geological time interval, based on which constructed diversity curve, and analyzed and interpreted the trends exhibited by the curve (Alroy, 2003). With the establishment of regional (eg, the Fossil Record File database or FRED data set of

New Zealand—Crampton et al., 2006; GSA Data Repository item 2011070, available online at www.geosociety.org/pubs/ft2011.htm—Rose et al., 2011; database of Neogene marine biota of tropical America or NMITA, http://eusmilia.geology.uiowa.edu/nmita.htm—Sessa et al., 2012; European Pollen Database, http://europeanpollendatabase.net/data/—Svenning et al., 2011; Neogene marine biota of tropical America or NMITA, http://eusmilia.geology.uiowa.edu/nmita.htm—Sessa et al., 2012), global (Sepkoski Jr., 2002; Paleobiology Database or PBDB, http://paleodb.org—Kiessling and Aberhan, 2007; Janus database, www.odp.tamu.edu/database, www.ngdc.noaa.gov/mgg/geology/drill.html—Powell and MacGregor, 2011; NEPTUNE, www.chronos.org and NEOMAP, http://www.ucmp.berkeley.edu/neomap), specific faunal (Coralline algae—International Fossil Algae Association at http://members.tripod.com/bruno.granier/index.html—Aguirre et al., 2000; Oligocene–Miocene Scleractinian coral occurrences in the circum-Mediterranean region—REEFCORAL database—Perrin and Bosellini, 2012), or floral (Climate Leaf Analysis Multivariate Program or CLAMP—http://clamp.ibcas.ac.cn—Fletcher et al., 2014; GSA Data Repository item 2012053, Table DR1—Cretaceous fossil wood occurrences and Table DR2—mean ring width measurements, www.geosociety.org/pubs/ft2012.htm—Peralta-Medina and Falcon-Lang, 2012) databases and programs and analyses of evolutionary, climatic, environmental, and other myriad varieties of geological and geobiological problems based on the open access provided by these databases, a variety of concerns arising out of biases were expressed. Alroy (2003) reviewed the problems associated with these and concluded that quantifying global diversity through developing a discipline-wide database is a multifaceted and scientifically important goal. Benton (2008) cautioned against the synonymy in using large databases as up to 40–50% errors are reported on genus and species level data on even a intensively revised and reviewed large clade like *Dinosauria*.

Given cognizance to these complexities, there are new computer programs being introduced as open-source software, with exclusive functions to analyze the biodiversity data with trait by trait character. For example, Clavel et al. (2015) recently developed a software named mvMORPH. This is a package of multivariate phylogenetic comparative methods for the R statistical environment and freely available on the CRAN package repository (http://cran.r-project.org/web/packages/

mvMORPH/). It allows fitting a range of multivariate evolutionary models under a maximum-likelihood criterion, can be extended to any biological data set with one or multiple covarying continuous traits and applied for fitting models with changes in the mode of evolution along the phylogeny. Thus, this program can analyze extinct taxa, for example, changes in evolutionary mode associated with global biotic/ abiotic events.

4.3 SPATIAL BIODIVERSITY

Analysis of the species *abundance* and *diversity* data in spatial domain, in the context of sea-level fluctuation is important in the light of the documented density of all size categories of organism's gradual decrease with increasing distance from the continental shelf and also due to the observation that species diversity shows a parabolic distribution with depth, reaching a peak in the bathyal zone, before decreasing to the abyssal plain (Rogers, 2000). Any change in the bathymetric variations may also shift the environmental boundaries and thus, impact the ecospaces too (Figure 3.2; Ramkumar, 2006 and Ramkumar and Sathish, 2009). Statistical analysis for the assessment of range determinants and prediction of species occurrence, or *Species Distribution Modeling* (SDM) is applied to mapping and understanding the distributions of taxonomic entities or ecological or ecosystem types. SDM offers new possibilities for estimating and studying past organism occurrence and distributions across space and/or time (Svenning et al., 2011). Bambach et al. (2007) and Bush et al. (2007) presented 216 theoretically possible categories of modes of life depicted in the $6 \times 6 \times 6$ grid defined based on topographical location relative to the sea floor, motility, and feeding mechanism. This method offers subjective description of ecospaces and quantitative analysis of variability on a geological time scale.

4.4 TRUE AND APPARENT BIODIVERSITY—SIGNALS AND CAUSES OF DISTORTION

Three interrelated factors, namely, the *absolute heterogeneity* of the fossil record (preservation is biased over space and time by varying sedimentologic and taphonomic parameters—Sigwart et al., 2014), the *gnostic heterogeneity* (which represents our uneven knowledge of the record), and the *analytical framework* mask the true biodiversity signals. Villier and Navarro (2004) stated that the variations in recorded diversity over time present a scrambled signal that is modulated by a large number of variables, namely, the potential of particular life forms to generate evolutionary innovations, external constraints induced by the environment in its broad sense, the heterogeneity of the fossil record, and the analytical artifacts due to sampling bias. A key question is how to characterize and quantify the separate input of any given factor in the overall diversity signal. In this section, the biases encountered during biodiversity analyses are presented.

Kiessling and Aberhan (2007) listed that the amount of available energy in a particular environment, namely, the energy in the form of solar radiation or temperature, nutrient availability, predation pressure, competition, and temporal and spatial variability of environmental parameters control the diversity, radiation, and extinction of biota. Based on these premises, these authors suggested that instead of interpreting the biodiversity curves *per se*, suitable considerations should be given to those environmental categories and other artifacts. As these conditions are not static and change continually on geological time scales, the consequent adaptations of species to this continuum result in diversity. And the inability to adopt leads to extinction and forms the trends in biodiversity curves. Rook et al. (2013) emphasized the difference in the relationship between biologic and sedimentary processes at work in the marine and nonmarine realms, and also the differences in the magnitude of the bias imposed by the rock record on the biologic patterns. Comparison of particular organism against others provides clues on environmental, climatic, paleogeographic preferences, and the successfulness of the organism. However, as the diversity signal is a composite function and is dependent on a large number of factors, these simplistic approaches of differentiations were found untenable.

Vecoli and Hérissé (2004) suggested removal/avoidance of taxonomic, spatial, preservational, and time averaging biasing effects before analyzing the diversity patterns. Although the potential effects of geological and anthropogenic biases on accurate taxon counts are known for long, it is only more recently that substantial efforts are being made to correct these biases (Newham et al., 2014). *Taxonomic biases* can be significant sources of errors especially in literature-based diversity analyses and emanate from the inconsistent adoption, in different studies, of morphological criteria on which fossil taxa are defined and identified. *Spatial and preservational biases* are due to areal and/or vertical variability in sampling intensity and in preservation potential of fossils, which in turn influences the taxonomic composition as well as the size of the fossil populations under analysis. Biases on diversity estimates are to be expected if sampling is unevenly distributed among different sedimentary facies, or if the effects of facies shifts along a time coordinate are not taken into account and corrected for. Large variations in sample time averaging (ie, the total amount of time which is represented in the analyzed sample) can affect micropaleontological samples because of different sedimentation rates, reworking, and bioturbation.

Knoll et al. (1979) stated that there is always a bias in diversity of plant fossils because plants are rarely preserved as complete organisms, and that most vascular plants live in areas of erosion rather than in areas of sediment accumulation. Raup (1972, 1976) observed a strong correlation between observed taxonomic diversity and the availability of sedimentary rocks. Raup (1972) pointed out that spatial effect on biodiversity estimates are greatest at the species level and decrease at higher taxonomic ranks.

4.5 METHODS OF BIAS CORRECTION

Many studies, including Peters and Foote (2001), Kiessling (2002), Bush et al. (2004), Villier and Navarro (2004), Crampton et al. (2006), Kiessling and Aberhan (2007), Hendy (2009), Monnet (2009), Ruban (2007, 2008, 2009), Servais et al. (2010), Cascales-Miñana et al. (2013), Butler et al. (2013), Newham et al. (2014), suggested to analyze biodiversity without lithologic (by including fossils of different lithological types), preservation (by including fossils of different environmental settings/habitats), sampling (by including fossils of different geographic locations; see Butler et al., 2013, for a detailed review), and collection (varying collection efforts among geological intervals—Mannion et al., 2013) biases and the use of statistical tests to affirm the elucidated biodiversity patterns.

Cascales-Miñana et al. (2013) provided a detailed account on the potential pitfalls that may thwart fruitful biodiversity analysis of plant fossil taxa. Counting both monophyletic and paraphyletic taxa is indeed misleading when estimating diversity (Villier and Navarro, 2004) as few of these consist of evolutionary information, while the others can be thought of as sampling artifacts (Villier and Navarro, 2004). Mannion et al. (2013) presented a recent review of studies on biases and appropriate method of paleobiodiversity computation and analysis. Diversity variations through time are sensitive to autocorrelation, which can obscure the influence of external factors. The effect of autocorrelation is generally removed by working with detrended signals. Newham et al. (2014) suggested the use of reconstructions of ancient diversity patterns by correcting observed taxon counts for sampling biases. The correction methods fall broadly into two classes: model-based approaches that first attempt to quantify the available record using sampling proxies, then construct models of expected diversity for use as a correction factor to the observed diversity; and sampling standardization (or subsampling) approaches that simulate an equal, or fair sample of ancient diversity among time intervals based directly on occurrence data, making fewer assumptions about drivers of the record. *Sampling standardization* is a method that reduces the differences in sampling intensity and rarefaction (Crampton et al., 2006). It is based on randomized resampling of lists. For the time intervals of interest, taxonomic occurrences or entire lists are sampled randomly without replacement and, to overcome the effects of stochastic resampling error, repeatedly. Kiessling (2002) cautioned about the influence of *beta diversity*, that is, the taxonomic differentiation between different localities and regions. The more different the faunas from different localities are, the higher is the beta diversity and the higher the global diversity will be. Hence, the measures such as computation of *Dice coefficient*, which gives double weight to presence data but is otherwise redundant with the more widely applied *Jaccard coefficient* is suggested. The index is suitable for incomplete and qualitative rather than quantitative faunal lists.

Hendy (2009) analyzed the biodiversity trends of molluscs and brachiopods in terms of lithification bias and concluded that lithified rocks preferentially destroyed the original diversity. Influence of local environmental factors creating artifact of biodiversity has been reported by Kiessling (2002). For example, the deeper marine deposits are likely to record a higher diversity than shallow marine deposits owing to intrinsically higher fossilization potential. Crampton et al. (2006) reviewed the previous studies that questioned the enhanced biodiversity during Cenozoic based on the well-documented New Zealand shelfal marine molluscan data after removing the potential taphonomic and systematic biases and methodological uncertainties, namely, non-uniform loss of aragonitic faunas, biostratigraphic range errors, taxonomic errors, choice of time bins, choice of analytical protocols, and taxonomic rank of analysis and concluded that no evidence was found for an increase in marine molluscan diversity through the Middle and Late Cenozoic.

Checks should be made to differentiate observed patterns with that of general nature of the time slice under study. For example, diversity decrease by lower origination rather than by higher extinction is a frequently observed pattern of marine diversity throughout the Phanerozoic (Bambach et al., 2004). Similarly, turnover values, a measure of the intensity of the changes of the whole community, may also be artificially increased by the presence of gaps in faunal successions. Comparative inter-ocean and across-time analyses of both living and fossil biota are necessary to extract better understanding of fundamental biodiversity and evolutionary patterns throughout Earth's history (Roy et al., 1998). Miller and Mao (1995) and Miller (1997) stated that biodiversity trajectories are dependent on geological processes and hence biotic diversification should be analyzed on the basis of paleogeographically and paleoenvironmentally homogeneous datasets rather than on global compendia.

Peters and Foote (2001) analyzed correlation of the normalized maximum and minimum diversities registered with the number of lithostratigraphic units per million years. To determine potential relationships and corresponding correlations between the diversity measures and the sampling proxies, linear regression analysis was performed by Cascales-Miñana et al. (2013). Butler et al. (2013) suggested that the conductance of exceptional sampling in a restricted spatio-temporal segment, such as *Lagerstätten*, can obscure the relationship between taxonomic richness and "background" sampling and hence, time series generalized least-squares multiple regression models to determine the relationships between richness and various combinations of the explanatory variables should be attempted.

The study of Cascales-Miñana et al. (2013) is one of the robust examples that analyzed different biases that influence biodiversity trends and suggested methods to overcome the problem. These authors evaluated the diversity fluctuations and the corresponding sampling biases, by measuring levels of taxonomic extinctions and by exploring disruptions to similarity patterns between time units. To analyze the direct consequences for identifying extinction processes from diminishing levels of diversity, different sampling proxies, such as the volume of sedimentary rocks, the area of sediment accumulation, the total (erosion-accumulation) area, the global sediment mass, and the number of geologic units through time, were employed. The results suggested that while studying diversity dynamics to quantify the real scale of extinction, the effect of sampling bias should be given due consideration. After testing 16 metrics in two different temporal frameworks, by using criteria like the adjustment between the descriptive extinction metric and the derived probabilistic profile, and by using normalized metrics that discounted short-lived taxa, these authors were able to improve interpretation of extinction intensity. Significant effect of sample size and abrupt reductions of similarity coefficients between successive time units were also observed and were eliminated by using clustering methods.

CHAPTER 5

Climate—Sea Level Coupling and its Influence on Biodiversity

5.1 PALEOBIODIVERSITY DYNAMICS IN THE LIGHT OF EUSTATIC SEA-LEVEL CYCLES

Though biodiversity could be influenced by other factors than sea-level change at least for certain fauna and during certain geochronological intervals (Rose et al., 2011), a parallelism between increased rates of tectonic activity, ecological changes, and patterns of radiations and extinctions suggest the existence of causative relations between internally driven physical modifications in the geosphere and biological changes (Vecoli and Hérissé, 2004), and the role of sea-level changes. Similarly, secular distribution of microbial carbonates due to the competitive exclusion by eukaryotes in general, and metazoans in particular, links major patterns of biodiversity over long time scales under the influences of a variety of factors that each changed through time (Riding, 2006). However, a recent succinct review (Holland, 2012) stated that the role of sea-level fluctuations over biodiversity is known to paleobiologists in the 1900s itself, but only recently quantitative evidences relating sea level, areal extents of habitat with biodiversity are being appreciated. Ruban (2010a) was unequivocal that *"a relationship between global sea levels and the diversity of marine invertebrates throughout the Phanerozoic remains an urgent matter for debates. Its recognition depends on a proper selection of diversity and eustatic curves. A comparison of changes in the revised sample-standardized generic diversity and long-term global sea level changes provides a weaker evidence for a direct covariation than established earlier, although the eustatic control on diversity dynamics of ∼74% of the Phanerozoic marine invertebrates was important."*

Smith et al. (2001) presented a detailed account on the environmental, ecological, and habitat-diversity changes that are associated and/or that may be expected as attendant to the sea-level changes. Many factors control the distribution and preservation potential of individual

Marine Paleobiodiversity. DOI: http://dx.doi.org/10.1016/B978-0-12-805415-4.00005-X

taxa, the most important being nutrient availability, substrate type and turbulence and all these three are directly or indirectly correlated with water depth (Smith et al., 2006). As sea-level fluctuations introduce bathymetric variations and alter the habitats, the extents of which are dependent on the amplitudes of fluctuations and the expansion/ contraction/creation/destruction of ecospaces are dependent on the geographic position of the habitat and the direction of sea-level fluctuation (rise/fall), the resultant biodiversity might reflect sea-level fluctuations. Masse and Fenerci-Masse (2011) documented four drowning events from platform carbonates of SE France that took place during Upper Barremian–Bedoulian and analyzed their significant role in the stratigraphic organization of shallow platform carbonates. Building on this initial documentation, Masse and Fenerci-Masse (2013) reported changes in trophic conditions and replacement of rudists by *Palorbitolina* communities, reduction of the calcification potential of benthic and planktonic organisms and a nannoplankton crisis as a direct consequence of platform drowning.

Relationship between relative sea-level fluctuation and paleobiodiversity was documented by many authors from a wide variety of environmental, paleogeographic, and chronologic domains. For example, Newell (1967), Hallam and Wignall (1999), Purdy (2008), and Melott and Bambach (2011) examined the relationships between fauna and sea-level fluctuations, O'Dogherty et al. (2000) and Sandoval et al. (2001a,b) demonstrated the dependence of the regional fossil diversity on sea-level changes, Ruban (2010a,b) documented the paleodiversity dynamics in the light of eustatic sea-level cycles, and Gale et al. (2000) analyzed the faunal diversity in the sequence stratigraphic framework and correlated the observed diversity trends with sediment geochemistry, paleoproductivity, etc., and found correlation of diversity dynamics with eustatic sea-level fluctuations.

Based on a large dataset comprising 90% of known ammonite genera (which are normally omitted in long-term biodiversity studies as most of these are *singletons*, ie, occur in only in a single stage), Yacobucci (2005) conducted time series analysis to find white noise, discounted the notion of self-organized criticality among ammonite originations and extinctions, and interpreted the white noise as the result of short-term evolutionary radiations principally triggered by physical factors such as sea-level cycles than biological factors. On the contrary, Holland (2012) demonstrated that the sea-level fluctuation and change in habitable area do not have a straightforward relationship in shallow marine settings and also that different areas respond differently to same magnitude of sea-level fluctuations. Vecoli and Hérissé (2004) are of the view that the acritarch diversity varies along an inshore—offshore transect as a function of varying paleoecologic conditions, including variation in water depth due to paleoenvironmentally controlled proportional changes in species composition of the assemblages. Martin et al. (2008) recorded evolution of phytoplankton stoichiometry related to tectonic (Wilson) cycles of supercontinent rifting and reassembly and associated climate change, broad sea-level rise, warm climate (high CO_2), sluggish ocean circulation, and anoxia.

A wide range of factors including global cooling, sea-level changes, volcanic events, climatic reversal, orogeny, superplume activity, and asteroid breakup and its consequent high meteorite flux have recently been postulated to have caused or at least promoted the Great Ordovician Biodiversification Event or GOBE (Servais et al., 2010), among which, climatic reversal and sea-level change were considered to have significant influence. Despite the absence of consensus (Hallam and Wignall, 1999; Ruban, 2008), eustatic fluctuations are often invoked among the most significant factors for the benthic macrofaunal diversity (Newell, 1967; Purdy, 2008), at least for some time intervals and particular fossil groups (McRoberts and Aberhan, 1997; Stanley, 2007). Neraudeau et al. (1997) reported higher echinoid diversity during sea-level highstands. Contrarily, Smith et al. (2006) reported greatest diversity in the transgressive system tracts of inshore settings and lowest in the highstand system tracts of outer shelf settings. The influence of sea level and climate over biodiversity has been exemplified by the study of Aguirre and Riding (2005) who have documented the diversity trends of dasycladalean green algae from Carboniferous to Pliocene. These authors have observed that except during Late Cretaceous, biodiversity broadly tracked global fluctuations in temperature and sea level.

5.2 BIODIVERSITY DYNAMICS DUE TO CLIMATIC AND OCEANOGRAPHIC CHANGES

Elevated pCO_2 in the atmosphere and resultant greenhouse effect warms the oceans and triggers a chain of reactions. Ocean warming acts such that warming increases the surface ocean stratification, which in turn affects the surface water light regime and nutrient input from deeper layers and impacts primary production. Tappan (1968) proposed a model for the evolution of pelagic systems which later became widely accepted. According to this model, biota are intimately related with continental physiography, sea-level changes, climatic changes, availability of nutrients, atmospheric oxygen and CO_2 levels, and consequent changes in the development of oceanic surface- and deep water masses (Monnet, 2009). A common causative mechanism in models predicting declines of global primary and export production is the increased stratification of the ocean in low to mid latitudes and a slowing of the thermohaline circulation, reducing nutrient availability in surface water layers. Oxygen transport to the deep sea by downwelling water masses will be weakened by freshening by increased meltwater input from glaciers, altering patterns of ocean mixing, slowing down the conveyor belt and leading to progressive depletion of the ocean's oxygen inventory which destabilizes ecosystems through altered food web dynamics, and reduces abundance of habitat-forming species (Hoegh-Guldberg and Bruno, 2010). This cascading effect significantly reduces the species populations and limits the potential for adaptation (Bijma et al., 2013). These two are the traits that become observable in biodiversity trends of the past.

While analyzing global paleoenvironmental perturbations on the evolution of organic-walled microphytoplankton based on calculation of evolutionary rates from latest Cambrian through Ordovician and development of a detailed diversity curve at specific level, Vecoli and Hérissé (2004) stated that the variations in acritarch diversity and origination/extinction rates were found to be related to the effect of the climatic shift from hot house to ice house conditions and the associated oceanic overturn on the microphytoplankton communities. Valentine (1968) commented that a global cooling trend, ie, sea-level fall, enforces a differentiation of global environments (*provincialism*) and accelerates diversity growth. Similarly, an isolation of islands and continents, which is related to eustatically driven transgressions, contributes to *endemism* (Ruban, 2010a).

Based on the study of z-corals of Oligocene−Miocene, in the Mediterranean region, Perrin and Bosellini (2012) opined that abundance of a species indicates that it is best equipped to diverse environmental conditions and hence can have a long persistence time and a wide distributional range. It results to higher probability of speciation. The patterns of taxonomical richness reflect the complex interplay of processes that acted over a wide range of spatial and temporal scales. Using marine isotope records of sulfur and strontium as continuous proxies for variability of igneous magmatism, Prokoph et al. (2013) were able to quantify the relationships between the events of large igneous provinces, changes in oceanic chemistry and marine biodiversity evolution since last 350 Ma. Harper (2006) stated that the profound changes in biodiversity and morphological disparity were matched by dramatic changes in the planet's marine ecosystems from the less well-organized associations of the Cambrian evolutionary faunas to the more structured paleocommunities and diverse assemblages of the Paleozoic evolutionary faunas. However, many Phanerozoic ecological events were decoupled from major changes in diversity, implying the controls and constraints on diversity trends and ecological structures were not necessarily similar. Ecological structures can be strongly influenced by the abundance or disappearance of key species rather than by mere increase or decrease of biodiversity.

Kammer and Ausich (2006) documented the enhancement of ecospace by extensive carbonate ramp development and associated open circulation. After the Late Devonian mass extinction, rapid radiation of crinoids during Mississippian occurred. It was brought to end by widespread glacial advancement, major drop in sea level, and disappearance of major epicontinental seas during the end Mississippian. This deprivation of ecospace has brought to an end of the age of the crinoids. As the epicontinental sea magnified the circulation effects, the chemistry of the sea water was influenced by onshore—offshore distance, source of sea water, and salinity gradient. These changes might have been influenced by the attendant environmental conditions as well. These observations directly link the climate—sea-level fluctuation—ecospace creation and biodiversity. Numerical paleoclimatic and paleoceanographic modeling applied to Mid-Cretaceous situations (Poulsen et al., 1998) have shown that even small modifications to continental geometry are potentially able to significantly alter global oceanic circulatory patterns with profound, global effects on the marine environment, which ultimately affect the biodiversity. Monnet (2009) analyzed the biodiversity patterns of ammonoids from three major areas (Europe, Tunisia, and the Western Interior) in order to better understand the relationships between ammonoid biodiversity patterns and abiotic factors during the

Cenomanian–Turonian interval. This study had revealed that the changes in ammonoid diversity are compatible with the exceptional high sea level occurring at that time and with concomitant regional climate changes. These inferences were corroborated with the fact that the Cenomanian–Turonian boundary recorded the highest sea level of the Mesozoic (Haq et al., 1988), highest atmospheric CO_2 concentrations, a global warming, a global oceanic anoxic event, and a global major positive $\delta^{13}C$ excursion coupled with massive deposition of organic-rich sediments (Monnet, 2009). It is further reinforced by the observations on the recent ocean warming, acidification and deoxygenation revealed a dramatic effect on the flora and fauna of the oceans with significant changes in distribution of populations, and decline of sensitive species (Bijma et al., 2013). On the contrary, applying these modern examples to older records might introduce potential pitfalls as exemplified by a study by Smith et al. (2001). On examination of Cenomanian/Turonian bioevent and associated features in Western Europe, Smith et al. (2001) commented that though the large-scale sea-level cycles and sea-level change could have altered the ratio of shallow-deep marine sedimentary records, the rock record and taphonomic megabias could have exacerbated the diversity trends and might explain the observed extinction and radiation events.

Heimhofer et al. (2005) documented the palynological records from the Western Portuguese and Algarve basins, and supplemented the biostratigraphic and sedimentologic evidences to demonstrate a major radiation phase of angiosperms during the Early Albian as a result of global scale major climatic and oceanographic perturbations. Based on the stepwise increments and latitudinal diachroneity, these authors interpreted a stepwise incremental diversification and successive migration that resulted in higher diversity and spatial distribution with progressive time, principally under major environmental changes in terms of tectonic events, several episodes of climatic warming and cooling, significant changes in humidity, hydrological cycling and weathering conditions, and release of excess greenhouse gases from igneous events.

Drowned coral reefs are the results of rapid pulses of relative sea-level rise or reduction of benthic growth by deterioration of the environment (Hallock, 2005), especially paleogeographic reorganization, tectonics, changes in ocean temperature, productivity, benthic carbonate producers, oversteepening and shelf erosion and burial by clastics, modifications in the atmospheric carbon dioxide levels, and methane hydrate dissociation event (Masse and Fenerci-Masse, 2013). Diversification of the marine biosphere is intimately linked to the evolution of the biogeochemical cycles of carbon, nutrients, and primary productivity. A meta-analysis of the ratio of carbon phosphorus buried in sedimentary rocks during the past 3 billion years by Martin et al. (2008) indicated that evolving food quantity and quality was primarily a function of broad tectonic cycles that influenced not just carbon burial but also nutrient availability and primary productivity. As these two are intimately linked to the oceanographic processes and the extents of habitats and bathymetry, relative roles of sea level in regulating habitat heterogeneity and biodiversity are inferred. Toshimitsu et al. (2003) built a large database of 790 ammonoid species from the Cretaceous of Japan and analyzed the species diversity through the 31 well-constrained Cretaceous substages of Japan to find 7 peaks of high diversity and 6 events of dwindling coinciding with the occurrences of oceanic anoxic events. Thus, this study unequivocally established the role of coupled climate—sea-level fluctuation and resultant nutrient variability over temporal scales that in turn caused the biodiversity trends to fluctuate.

Based on the relative abundance data from Early, Middle, and Late Permian silicified fossil assemblages of offshore shelf carbonate environments, significant increase of molluscs in relative abundance, from 0.8% in the Middle Permian (Guadalupian) to 65.4% in the Late Permian (Lopingian), was observed by Clapham and Bottjer (2007). These authors also inferred a significant ecological shift and environmental variability in proximity to the deep water mass which in turn may have given eurytopic molluscs a competitive advantage over more stenohaline brachiopods. These authors further suggested the prevalence of environmental control over species diversity and decoupling of ecosystem response between brachiopods and gastropods. This inference has serious implication for those paleobiodiversity trends based on polyphyletic or fossil communities of diverse organism groups, as a single process/event may invite/trigger diverse response from different groups. Schmiedl et al. (2010) analyzed the spatio-temporal variability of environmental parameters, biogeochemical processes, and resultant ecosystem dynamics. Based on these, the authors have demonstrated that the oxygenation and food availability of the deep sea ecosystems were influenced by orbital and suborbital climate variations of the high northern latitudes and the African monsoon system. The basic mechanism linking orbital forcing and monsoonal climate response is well documented.

5.3 ENVIRONMENTAL CHANGES AND RELATIVE INFLUENCES ON BIODIVERSITY AND HABITAT

According to Perrin and Bosellini (2012), *taxonomical richness* in any given region is the result of interaction between the three parameters, namely, recruitment of new taxa resulting from immigration of species from external areas, intraregional speciation either through allopatric or sympatric processes, and sustenance of already existing species within the considered region. In one of the systematic documentation and analysis of biodiversity as a function of habitat dynamics influenced by sea-level cycles, Smith et al. (2006) recorded about 3500 individual occurrences of echinoids and correlated them against a sequence stratigraphic framework representing onshore, mid-shelf, and deeper shelf habitats. Their study had demonstrated the prevalent marked differences in the composition and diversity of faunas both across the shelf at a single time interval and through time at the same locality, driven primarily by sedimentary facies changes, which were controlled by sea-level fluctuations. The study had also documented that the ranges of individual taxa expanded and contracted across the shelf as sea levels change. In mid-shelf environments, more onshore taxa appeared only near sequence bases, at times of lowest sea level, while those from more outer shelf settings were found during highstand intervals and created a cyclic pattern of diversity.

Based on global diversity patterns of Late Jurassic–Early Cretaceous (Tithonian–Berriasian) radiolarian faunas, Kiessling (2002) reported the occurrence of a significant diversity gradient from tropical–subtropical to subpolar latitudes and an asymmetry of biodiversity between southern and northern hemispheres. He had concluded that latitudinal diversity gradient to a time-invariant feature of the Earth possibly driven by gradients of biome size. Rogers (2000) observed the occurrences of the most deep sea benthic biota species-rich areas on the continental margins between 500 and 2500 m, which coincided with the present oxygen minimum in the world's oceans. He had concluded that the geographic and vertical distribution of many species is restricted by the presence of oxygen-minimum zones. As the cycles of global warming and cooling (and thereby resultant sea-level rise and fall) lead to periods of expansion and contraction of oxygen-minimum layers throughout the world's oceans, such shifts in the global distribution of oxygen-minimum zones might influence the habitats and thereby the species abundance and diversity. This feature apparently explains the habitat and biodiversity changes on a spatio-temporal scale. There are observed trends for some species that shifted their principal dwelling/inhabiting areas poleward and into deeper, cooler waters. However, range shifts within short time frames may be unlikely for many species, such as long-lived, slow growing, sessile habitat-forming species, may be exposed to increased extinction risk. In the case of coastal species, a poleward shift in distribution may be limited by geography as organisms simply *run out* of coastline to migrate along and are faced with a major oceanic barrier to dispersal (Bijma et al., 2013). The unusual highstand of Albian—Cenomanian induced a shift toward deeper water environments during the Late Cenomanian caused a reduced partitioning of epicontinental seas. It also induced a more uniform climate with a drastically reduced seasonal contrast of sea surface temperature in epicontinental seas as well as a reduction of the equator-to-pole sea surface temperature gradient. The reduced partitioning of epicontinental seas and the more equable climate contributed to the homogenization of marine habitats during the Late Cenomanian, which in turn caused a decreasing endemism and decreasing species richness. The decreasing provincialism of ammonoids observed during the Late Cenomanian probably resulted from this homogenization of marine habitats (Monnet, 2009).

Crampton et al. (2006) are of the view that though the area of habitat *per se* may not have direct control over biodiversity, but richness is the interplay between tectonism, nutrient, oceanographic parameters, environmental stress, etc., together with habitat. In addition to forming an ecosystem and habitat of their own, shallow water reefs produce complex structures that support high-diversity communities (Hallock, 2005), provide a variety of microhabitats and ecosystem services to a variety of organisms and thus stand testimony to the impact of sea-level fluctuations over habitat and biodiversity dynamics. While analyzing the Phanerozoic evolutionary trends of brachiopods, Ruban (2009) observed no direct relationships between intensity of turnovers and eustatic fluctuations. However, the changes in the diversity structure recorded with the *Gower Index* provided evidence that eustatic lowstands were more favorable for intensification of these changes. It was one of the pioneering and conclusive quantitative evidences of impact of sea-level variations over biodiversity as a function of deprivation of ecospace. As noted by Bijma et al. (2013), the fluctuations impact the benthic species severely, as they cannot shift to newer niches in tune with the rapidly shifting habitats.

CHAPTER 6

Impacts of Sea-Level Fluctuations on Biotic Turnover

In an editorial, Chen et al. (2014) noted that climate extremes and associated environmental stresses may become a characteristic of our future. Earth's ecosystems have suffered severe environmental stress and dramatic climate change throughout the geological history. How rapidly ecosystems can adjust to abrupt climate change and environmental extreme is a fundamental question accompanying present-day global warming. An important tool to address this question is to document and understand the equivalents strewn as mass extinction events in the geological record, in particular where the magnitude and/or rates of change in the global climate/environment system impacted the biodiversity. Thus, the sedimentary and paleontologic record of large-scale ecosystem collapse should provide insight into the potential response of ecosystems to present-day climate and environmental change. In this chapter, studies on events of biotic turnover, radiation, and extinction as a result of sea-level fluctuations and associated environmental changes are presented.

6.1 LONG-TERM CHANGES AND BIODIVERSITY DYNAMICS

The contrast between carbonate platforms dominated by microbial deposits in the Precambrian and those dominated by algal and invertebrate skeletons and their debris in the Phanerozoic is essentially due to the long-term changes in the processes of carbonate sediment accumulation (Riding, 2006) and as the carbonates are largely bioproduced, long-term changes in biotic, particularly the skeletal producers and their preservational conditions can be inferred. During Phanerozoic evolution, life suffered at least "big five" or first-order mass extinctions, namely, during the end-Ordovician, end-Frasnian, end-Permian, end-Triassic, and end-Cretaceous, respectively (Sepkoski, 1981) and other 15 minor mass extinctions. Of the five first-order mass

Marine Paleobiodiversity. DOI: http://dx.doi.org/10.1016/B978-0-12-805415-4.00006-1

extinctions, the end-Ordovician (Hirnantian) event was associated with the climatic and oceanographic effects enforced by the sea-level fluctuation (Harper et al., 2014). Markov and Korotayev (2007), on examination of the Phanerozoic biodiversity pattern, commented that it follows a hyperbolic shape meaning, a feedback between the diversity and community structure complexity. Owing to this complexity in the community structure and its influence on biodiversity growth (as the only limiting factor of diversity growth of any taxa is its ability to occupy new ecospace—Benton, 1995), any destabilizing factor, when introduced, could be catastrophic, leading to extinction.

Smith et al. (2006) suggested examining biodiversity in response to sea-level fluctuations such that deeper water species should appear first in offshore settings and migrate into more onshore settings when sea levels have risen. Similarly, one might expect to see shallow water taxa migrating into more outer shelf settings during times of lowest sea-level stand. These stands confirmed from the study of Powers and Bottjer (2009) who have documented elevated bryozoan extinction rates during the Late Permian and Late Triassic that were coupled with major changes in their habitats from nearshore to mid-shelf settings. Bryozoans gradually disappeared from deep-water offshore settings during the Late Permian and from nearshore and offshore settings during the Late Triassic. It essentially enforced a taxonomic switch between stenolaemate and gymnolaemate bryozoans and a long-term ecological change. The importance of this long-term extinction-related stress on marine settings and organisms reinforces the notion of an oceanographic process as the trigger of both the Late Permian and Late Triassic extinctions (Powers and Bottjer 2009). Ruban (2010b) recognized four Paleozoic major radiations namely, during the Early Cambrian, Ordovician, Early Devonian, and Mid-Permian in marine realm and found these events correlated with major sea-level fluctuations. He reasoned that when global sea level rose with low magnitude fluctuations and huge well-connected oceans provided a peculiar space for biological innovations and when sea level fell the presence of elongated island chains enhanced the radiation by a rapid dispersal of organisms across shallow oceans and along connected shore lines. Hallam (1977) also found that eustatic changes significantly controlled the global bivalve diversity.

The Mesozoic is a crucial time in earth history for investigating the origin, diversification, and evolutionary process (Wang and Axsmith

2008). Servais et al. (2010) stated that two sustained rises in the biodiversity of marine organisms took place during Earth history. The first is recorded during the Early Paleozoic and the second started at the beginning of the Mesozoic. Ruban (2007) reported the rises of ammonite diversity with transgressive episodes and falls to regressions which are superimposed on a long-term, slight species decline. He has also recorded a correspondence between eustatic rises in the Early Jurassic and during the Bajocian-Callovian with significant radiations of bivalves, and a peak bivalve diversity during Late Jurassic when sea level was highest and a decrease of bivalve diversity during Tithonian coinciding with eustatic fall. He has also compared the global belemnite diversity changes and sea-level fluctuations to find all three increases in diversity of belemnites during Pliensbachian–Toarcian, Bajocian, and Callovian–Oxfordian, corresponded to global eustatic rises. Albian–Cenomanian interval is characterized by a global second-order transgressive trend. This unusual transgression peaked in the Early Turonian which records the highest highstand of the Mesozoic and this significant high sea level resulted in major environmental changes and influenced diversity patterns of ammonoids during the Late Cenomanian (Monnet, 2009).

6.2 BIODIVERSITY DYNAMICS DUE TO SHORT-TERM SEA-LEVEL CHANGES AND PERTURBATIONS

The fossil record is inherently biased by factors that distort perceptibly secular patterns of past biodiversity and ecological change (Hendy, 2009) within which apparently recognizable perturbations in terms of sudden and/or stepwise or long-term extinction events are embedded. Few of these events are ascribed to sea-level fluctuations. Analyses of the taxonomic severity of biodiversity crises have a long history (see Benton, 1999, for a detailed review) which led to the recognition of big 5 and 15 less severe extinction events in the Phanerozoic. McGhee Jr et al. (2004) are of the opinion that it is important to analyze the ecological impact of biotic crises, rather than the magnitude of the biodiversity loss. There are instances of decoupling between them (McGhee Jr et al., 2013) as well. However, it is difficult to assess the impact on ecosystem as no system is available to measure ecological severity on a ratio scale and a method of ranking the *ecological severity*. Accordingly, these authors provided a scale of category, namely, I, II,

IIa, and IIb, representing existent ecosystems collapse. These severity classes are described in terms of:

i. replaced by new ecosystems post-extinction,
ii. existent ecosystems disrupted, but recover and are not replaced post-extinction,
iii. disruption produces permanent loss of major ecosystem components and temporary disruption,
iv. pre-extinction ecosystem organization reestablished post-extinction in new clades, respectively.

In addition, these authors suggested that the rank system is applicable to marine and terrestrial biomes and ecosystems. Based on a 52 Ma carbon isotopic reference curve generated for Comanche Platform, Texas, Phelps et al. (2015) recognized four stages (namely, equilibrium, crisis, anoxic, and recovery) of platform and associated changing patterns of facies, fauna, and relative sea-level trends that developed in response to global carbon-cycle perturbations.

Perrier et al. (2012) studied the impact of two Ordovician ash falls of different intensities to determine the recovery patterns of benthic ostracods and reported that the postcrisis ecosystem showed strong perturbations of diversity and abundance for a long period of time and significant ash falls led to marked rearrangement of assemblages and extinction of some taxa while less prominent volcanic episodes only resulted in temporary changes in the assemblage structure. McGhee Jr et al. (2004) are of the opinion that the environmental degradation produced by the end-Ordovician glaciations resulted in a loss of marine diversity much greater than that of the end-Cretaceous crisis. Major biological crises are attributed to a variety of factors, whose respective contributions and interactions are often difficult to disentangle. During the Late Barremian–Bedoulian, drastic changes recorded in the pelagic record include, Oceanic Anoxic Event, modification of the isotopic composition of the global carbon pool, volcanic degassing and release of methane hydrates contained in the marine sediments and resulted in a rise in atmospheric carbon dioxide pressure and, as a consequence, reduction of the calcification potential of benthic and planktonic organisms and a nannoplankton crisis (Masse and Fenerci Masse, 2013). The Cenomanian–Turonian boundary was a period of high organic matter burial coupled with a high positive excursion in the carbon isotope record. This event has been widely recognized to reflect anoxic conditions and is concomitant with one of the major mass extinctions of the Phanerozoic. It has been considered to be a typical example of global extinction caused by the spreading of anoxic waters (Monnet, 2009). Examination of the parallel diversity trends of total prasinophytes, total chlorophytes, total acritarchs, and total phytoplankton led Hérissé et al. (2009) to interpret Late Gorstian environmental changes and an upward decline in total phytoplankton diversity as the result of shallowing marine conditions and the emergence of shoreline carbonates. Progressive continental reshuffling and associated changes in oceanic circulation played a large role in disrupting biotic distributions during Late Mesozoic and Early Cenozoic (Stilwell, 2003). In a recent report, Vandenbroucke et al. (2015) stated that the metal poisoning was a contributing kill mechanism that operated during significant change in ocean basin volume, especially during the Ordovician–Silurian paleobiological event.

McGhee Jr et al. (2004) are of the view based on the analysis of five Phanerozoic biodiversity crises by ecological severity that the taxonomic and ecological severities of the events are decoupled and that selective elimination of dominant and/or keystone taxa that occurs in the ecologically most devastating biodiversity crises indicate a strategy emphasizing the preservation of taxa with high ecological value. These authors have reported the evidence for the decoupling is the end-Cretaceous biodiversity crisis, which is the least severe in terms of taxonomic diversity loss yet is ecologically the second most severe event in the entire Phanerozoic the end-Ordovician biodiversity crisis. The environmental degradation produced by the end-Ordovician glaciations precipitated a major loss of marine diversity, yet the extinction failed to eliminate any key taxa or evolutionary traits, and was of minimal ecological impact. The most easily identified proxy of high biotic stress conditions is the reduction in species diversity or species richness. The Late Maastrichtian mass extinction provides an excellent example to illustrate the progressive nature of biotic stress and its effect on biodiversity (Keller and Abramovich, 2009). Based on the study of sedimentary sequences analyzed from Tunisia, Egypt, Texas, Argentina, and the South Atlantic and the Indian ocean regions, Keller and Abramovich (2009) revealed that the biotic stress response appears uniform, regardless of the cause, varying only with the degree of biotic stress induced by oxygen, salinity, temperature and nutrient variations as a result of climate and sea-level changes and volcanism which caused the same detrimental biotic effects as observed in the aftermath of the Cretaceous/Paleogene (K/Pg) mass extinction, including the disappearance of most species and blooms of the disaster opportunists. Fara (2004) reported the survival of virtually all lissamphibian lineages at the K/Pg. This survival pattern is consistent across analyses made at various temporal, geographic, and taxonomic scales. These traits made Fara (2004) to conclude that long-term climatic disruption is not a viable hypothesis for the K/Pg extinctions. On the contrary, based on a first ever survey on the entire Paleocene record of the Southern Hemisphere, comprising 515 recorded molluscan taxa, Stilwell (2003) commented that the diversity of bivalves and gastropods across the K/Pg boundary stems from a rapid, evolutionary burst of speciation in the Danian for gastropods, especially carnivorous forms with planktotrophic development, which filled the ecospace vacated by the extinction. In the Southern Hemisphere fossil record, deposit feeders were less affected by the extinction event and seemingly more extinction resistant, but other important factors related to stratigraphic range and spectrum of life habits/habitats affected survivorship success. Suspension feeders, especially epifaunal forms, were hit hard by the extinction, but bounced back within a few million years by Late Danian at the latest, but at lower diversity than during the Late Cretaceous. Coxall et al. (2006) demonstrated that the planktonic foraminifera that were severely affected by the end-Cretaceous extinction event were able to recover only after the restoration of oligotrophic oceans and oceanic carbon flux. Sessa et al. (2012) interpreted that K/Pg extinction caused significant restructuring of the ecological composition of offshore assemblages and also that the ecological effects were facies specific, with offshore faunas displaying dramatic reorganization and shallow subtidal biotas remaining relatively unchanged across the extinction. Based on the observed lack of latitudinal pattern in mammal species diversity during Paleocene, Rose et al. (2011) concluded that the Paleogene extinction event could have disrupted climate and ecosystems in a way that led to unstable mammal communities during the Paleocene.

Based on a review of extinction, recovery and survival patterns of foraminifera across K/Pg to Eocene/Oligocene boundaries, Molina (2015) proposed new definitions for characterizing the extinction events. Accordingly, *sudden mass extinction* characterizes virtually instantaneous disappearance of organism(s) and the process would have taken a few years or decades (e.g., Cretaceous/Paleogene boundary), *rapid mass extinction* occurs in relatively short events, around 100 ka (e.g., Paleocene/Eocene and Eocene/Oligocene boundaries), *slow mass extinctions* are suggested to last around 1 to several Ma (e.g., Bartonian/Priabonian transition).

CHAPTER 7

What Holds for the Future

Though the sequence cycles of various orders and resultant changes in reservoir properties influenced by depositional—erosional regimes, varied responses of siliciclastic, carbonate, and mixed siliciclastic systems to the relative sea-level fluctuations resulted by tectano-eustasy, subsidence-uplift, sediment influx and climate are studied and published extensively, the impacts and responses of ecosystem and biodiversity patterns, despite being in common knowledge, have largely been overlooked and require detailed studies and adequate publication for better understanding of the past and reliable predictions for the future.

Relative sea-level fluctuations, either eustatic or introduced by any other cause, either long term or short term or due to perturbations as a result of endogenic or exogenic processes, do exercise corresponding changes on the biotic system either directly by creating or depriving or expanding or contracting habitats on oceans as well as onland, and/or indirectly through atmospheric, lithospheric, and oceanographic processes due to their coupled nature. The results are explicit in terms of radiation and extinction events as well as in terms of abundance, dwindling, colonizing, acquiring new morphologic traits of species and habitat heterogeneity.

Establishment of large databases at regional, global, faunal, and habitat-specific natures, provision of open access to them, the introduction of application of numeric methods of diversity data on a variety of scales and types, and the developments in removing biases have all provided the unprecedented opportunity to the geologists, or for the matter, paleobiologists to document and analyze the spatio-temporal diversity trends engrained in the geologic past for a better understanding. As reviewed in previous chapters, this better understanding has important and crucial implications for the sustenance of ecosystem and ecosystem services and functions, especially in the light of ongoing global warming.

Marine Paleobiodiversity. DOI: http://dx.doi.org/10.1016/B978-0-12-805415-4.00007-3

As the previous chapters have shown, despite being in knowledge the relative influence of sea-level fluctuations of all the scales and causes over the biodiversity dynamics, there is reluctance or to put it in perspective, neglect of documentation of biodiversity dynamics in response to the changes in relative sea level. Added to this is the decoupling between ecosystem and certain biota, at least during certain time slices and for few organisms due to complex interrelationships between intrinsic and extrinsic causes and processes, which remains enigmatic.

It requires a meeting of minds, convergence of ideas, and concerted efforts to foster joint and collaborative research by sedimentologists, paleobiologists, geochemists, stratigraphers, mineralogists, and researchers of other subdisciplines of geology to fill this vacant "ecospace" of documentation of habitat and biodiversity dynamics as a function/response of sea-level cycles. The "UNESCO IGCP-609 Climate-environmental deteriorations during greenhouse phases: Causes and consequences of short-term Cretaceous sea level changes" is an ideal platform and opportunity to initiate studies along these lines and it is hoped that this "fertile and luxuriant ecospace" will be colonized by the natural inhabitants (geoscientists) and make this habitat populated and diverse and save it from Lilliput effect and extinction.

REFERENCES

Aguirre, J., Riding, R., 2005. Dasycladalean algal biodiversity compared with global variations in temperature and sea level over the past 350 Myr. Palaios 20, 581–588.

Aguirre, J., Riding, R., Braga, J.C., 2000. Diversity of coralline red algae: origination and extinction patterns from the Early Cretaceous to the Pleistocene. Paleobiology 26, 651–667.

Alroy, J., 2003. Global databases will yield reliable measures of global biodiversity. Paleobiology 29, 26–29.

Bambach, R.K., Knoll, A.H., Wang, S.C., 2004. Origination, extinction, and mass depletions of marine diversity. Paleobiology 30, 522–542.

Bambach, R.K., Bush, A.M., Erwin, D.H., 2007. Autecology and the filling of ecospace: key metazoan radiations. Palaeontology 50, 1–22.

Benton, M.J., 1995. Diversification and extinction in the history of life. Science 268, 52–58.

Benton, M.J., 1999. The history of life: large databases in palaeontology. In: Harper, D.A.T. (Ed.), Numerical Palaeobiology. John Wiley, London, pp. 249–283.

Benton, M.J., 2008. How to find a dinosaur, and the role of synonymy in biodiversity studies. Palaeobiology 34, 516–533.

Bijma, J., Pörtner, H.-O., Yesson, C., Rogers, A.D., 2013. Climate change and the oceans—What does the future hold? Mar. Pollut. Bull. 74, 495–505.

Bush, A.M., Markey, M.J., Marshall, C.R., 2004. Removing bias from diversity curves: the effects of spatially organized biodiversity on sampling-standardization. Paleobiology 30, 666–686.

Bush, A.M., Bambach, R.K., Daley, G.M., 2007. Changes in theoretical ecospace utilization in marine fossil assemblages between the mid-Paleozoic and late Cenozoic. Paleobiology 33, 76–97.

Butler, R.J., Benson, R.B.J., Barrett, P.M., 2013. Pterosaur diversity: untangling the influence of sampling biases, Lagerstätten, and genuine biodiversity signals. Palaeogeogr. Palaeoclimatol. Palaeoecol. 372, 78–87.

Carter, R.M., Abbott, S.T., Fulthorpe, C.S., 1991. Application of global sea level and sequence stratigraphic models on Southern hemisphere Neogene strata from New Zealand. Int. Assoc. Sedimentologists Spec. Publ. 12, 41–65.

Cascales-Miñana, B., Cleal, C.J., Diez, J.B., 2013. What is the best way to measure extinction? A reflection from the palaeobotanical record. Earth Sci. Rev. 124, 126–147.

Chao, A., 1984. Nonparametric estimation of the number of classes in a population. Scand. J. Stat. 11, 265–270.

Chen, Z.-Q., Joachimski, M., Montañez, I., Isbell, J., 2014. Deep time climatic and environmental extremes and ecosystem response: an introduction. Gondwana Res. 25, 1289–1293.

Cherns, L., Wheeley, J.R., Wright, V.P., 2008. Taphonomic windows and molluscan preservation. Palaeogeogr. Palaeoclimatol. Palaeoecol. 270, 220–229.

Clapham, M.E., Bottjer, D.J., 2007. Permian marine paleoecology and its implications for large-scale decoupling of brachiopod and bivalve abundance and diversity during the Lopingian (Late Permian). Palaeogeogr. Palaeoclimatol. Palaeoecol. 249, 283–301.

Clavel, J., Escarguel, G., Merceron, G., 2015. mvMORPH: anR package for fitting multivariate evolutionary models to morphometric data. Methods Ecol. Evol.. Available from: http://dx.doi. org/10.1111/2041-210X.12420.

Coxall, H.K., D'Hondt, S., Zachos, J.C., 2006. Pelagic evolution and environmental recovery after the Cretaceous–Paleogene mass extinction. Geology 34, 297–300.

Crampton, J.S., Foote, M., Cooper, R.A., Beu, A.G., Maxwell, P.A., Cooper, R.A., et al., 2006. The ark was full! Constant to declining Cenozoic shallow marine biodiversity on an isolated mid-latitude continent. Paleobiology 32, 509–532.

Crampton, J.S., Foote, M., Cooper, R.A., Beu, A.G., Peters, S.E., 2011. The fossil record and spatial structuring of environments and biodiversity in the Cenozoic of New Zealand. Geol. Soc. Spec. Pub. 358, 105–122.

De Baets, K., Klug, C., Korn, D., Landman, N.H., 2012. Early evolutionary trends in ammonoid embryonic development. Evolution 66, 1788–1806.

De Baets, K., Klug, C., Monnet, C., 2013. Intraspecific variability through ontogeny in early ammonoids. Paleobiology 39, 75–94.

Fara, E., 2004. Estimating minimum global species diversity for groups with a poor fossil record: a case study of Late Jurassic–Eocene lissamphibians. Palaeogeogr. Palaeoclimatol. Palaeoecol. 207, 59–82.

Fioroni, C., Villa, G., Persico, D., Jovane, L., 2015. Middle Eocene–Lower Oligocene calcareous nannofossil biostratigraphy and paleoceanographic implications from Site 711 (equatorial Indian Ocean). Mar. Micropaleontol. 118, 50–62.

Fletcher, T.L., Greenwood, D.R., Moss, P.T., Salisbury, S.W., 2014. Paleoclimate of the late Cretaceous (Cenomanian–Turonian) portion of the Winton Formation, central-western Queensland, Australia: new observations based on CLAMP and bioclimatic analysis. Palaios 29, 121–128.

Foote, M., Sepkoski, J.J., 1999. Absolute measures of the completeness of the fossil record. Nature 398, 415–417.

Gale, A.S., Smith, A.B., Monks, N.E.A., Young, J.A., Howard, A., Wray, D.S., et al., 2000. Marine biodiversity through the Late Cenomanian–Early Turonian: palaeoceanographic controls and sequence stratigraphic biases. J. Geol. Soc. 157, 745–757.

Gale, A.S., Hardenbol, J., Hathway, B., Kennedy, W.J., Young, J.R., Phansalkar, V., 2002. Global correlation of Cenomanian (Upper Cretaceous) sequences: evidence for Milankovitch control on sea level. Geology 30, 291–294.

Gjelberg, J., Steel, R.J., 1995. Helvetiafjellet Formation (Baremian–Aptian) Spitsberen: characteristics of a transgressive succession. NPF Spec. Publ. 5, 571–593.

Goldhammer, R.K., Oswald, E.J., Dunn, P.A., 1991. Hierarchy of stratigraphic forcing: example from Middle Pennsylvanian shelf carbonates of the Paradox basin. In: Franseen, E.K., Watney, W.L., Kendall, C.G.St.C. (Eds.), Sedimentary Modeling, vol. 233. Kansas Geological Survey Bulletin, pp. 361–413.

Grammer, G.M., Eberli, G.P., Van Buchem, F.S.P., 1996. Application of high resolution sequence stratigraphy to evaluate lateral variability in outcrop and subsurface—Desert Creek and Ismay intervals, Paradox basin. In: Longman, M.W., Sonnelfeld, M.D. (Eds.), Paleozoic Systems of the Rocky Mountain Region, Rocky Mountain Section. Society of Economic Palaeontologist and Mineralogists, Texas, pp. 235–266.

Halfar, J., Zack, T., Kronz, A., Zachos, J.C., 2000. Growth and high-resolution paleoenvironmental signals rhodoliths (coralline red algae): a new biogenic archive. J. Geophys. Res. 105, 22107–22116.

Hallam, A., 1977. Jurassic bivalve biogeography. Paleobiology 3, 58–73.

Hallam, A., Wignall, P.B., 1999. Mass extinctions and sea-level changes. Earth Sci. Rev. 48, 217–250.

Hallock, P., 2005. Global change and modern coral reefs: new opportunities to understand shallow-water carbonate depositional processes. Sediment. Geol. 175, 19–33.

Haq, B.U., 2014. Cretaceous eustasy revisited. Global Planet. Change 113, 44–58.

Haq, B.U., Hardenbol, J., Vail, P.R., 1987. Chronology of fluctuating sea levels since the Triassic. Science 235, 1156–1167.

Haq, B.U., Hardenbol, J., Vail, P.R., 1988. Mesozoic and Cenozoic chronostratigraphy and cycles of sea level change. In: Wilgus, C.K., Hastings, B.S., Posamentier, H., Van Wagoner, J., Ross, C.A., Kendall, C.G.St.C. (Eds.), Sea Level Changes, vol. 42. Society of Economic Palaeontologists and Mineralogists, Texas, pp. 71–108.

Harper, D.A.T., 2006. The Ordovician biodiversification: setting an agendafor marine life. Palaeogeogr. Palaeoclimatol. Palaeoecol. 232, 148–166.

Harper, D.A.T., Hammarlund, E.U., Rasmussen, C.M.Ø., 2014. End-Ordovician extinctions: a coincidence of causes. Gondwana Res. 25, 1294–1307.

Hays, J.D., Imbrie, J., Shackleton, N.J., 1976. Variations in the earth's orbit: pacemaker of the ice ages. Science 194, 1121–1132.

Heimhofer, U., Hochuli, P.A., Burla, S., Dinis, J.M.L., Weissert, H., 2005. Timing of Early Cretaceous angiosperm diversification and possible links to major paleoenvironmental change. Geology 33, 141–144.

Hendy, A.J.W., 2009. The influence of lithification on Cenozoic marine biodiversity trends. Paleobiology 35, 51–62.

Hérissé, A.L., Mullins, G.L., Dorning, K.J., Wickander, R., 2009. Global patterns of organic-walled phytoplankton biodiversity during the late Silurian to earliest Devonian. Palynology 33, 25–75.

Hoegh-Guldberg, O., Bruno, J., 2010. The impact of climate change on the world's marine ecosystems. Science 328, 1523–1528.

Holland, S.M., 2012. Sea level change and the area of shallow-marine habitat: implications for marine biodiversity. Paleobiology 38, 205–217.

Hurlbert, S.H., 1971. The non-concept of species diversity: a critique and alternative parameters. Ecology 52, 577–586.

Jakubowicz, M., Berkowski, B., López Correa, M., Jarochowska, E., Joachimski, M., Belka, Z., 2015. Stable isotope signatures of Middle Palaeozoic Ahermatypic Rugose Corals—deciphering secondary alteration, vital fractionation effects, and palaeoecological implications. PLoS One 10. Available from: http://dx.doi.org/10.1371/journal.pone.0136289.

Kammer, T.W., Ausich, W.I., 2006. The "Age of Crinoids": a Mississippian biodiversity spike coincident with widespread carbonate ramps. Palaios 21, 238–248.

Keller, G., Abramovich, S., 2009. Lilliput effect in late Maastrichtian planktic foraminifera: response to environmental stress. Palaeogeogr. Palaeoclimatol. Palaeoecol. 284, 47–62.

Kiessling, W., 2002. Radiolarian diversity patterns in the latest Jurassic–earliest Cretaceous. Palaeogeogr. Palaeoclimatol. Palaeoecol. 187, 179–206.

Kiessling, W., Aberhan, M., 2007. Environmental determinants of marine benthic biodiversity dynamics through Triassic–Jurassic time. Paleobiology 33, 414–434.

Klompmaker, A.A., Ortiz, J.D., Wells, N.A., 2013. How to explain a decapod crustacean diversity hotspot in a mid-Cretaceous coral reef. Palaeogeogr. Palaeoclimatol. Palaeoecol. 374, 256–273.

Knoll, A.H., Niklas, A., Tiffney, B.H., 1979. Phanerozoic land-plant diversity in North America. Science 206, 1400–1402.

Laferriere, A.P., Hattin, D.E., Archer, A.W., 1987. Effects of climate, tectonics, and sea level changes on rhythmic bedding patterns in the Niobrara Formation (Upper Cretaceous), U.S. Western Interior. Geology 15, 233–236.

Laffont, R., Firmat, C., Alibert, P., David, B., Montuire, S., Saucède, T., 2011. Biodiversity and evolution in the light of morphometrics: from patterns to processes. C. R. Palevol 10, 133–142.

Levinton, J., 1988. Genetics, Paleontology and Macroevolution. Cambridge University Press, Cambridge.

Mannion, P.D., Benson, R.B.J., Butler, R.J., 2013. Vertebrate palaeobiodiversity patterns and the impact of sampling bias. Palaeogeogr. Palaeoclimatol. Palaeoecol. 372, 1–4.

Markov, A.V., Korotayev, A.V., 2007. Phanerozoic marine biodiversity follows a hyperbolic trend. Palaeoworld 16, 311–318.

Martin, R.E., Quigg, A., Podkovyrov, V., 2008. Marine biodiversification in response to evolving phytoplankton stoichiometry. Palaeogeogr. Palaeoclimatol. Palaeoecol. 258, 277–291.

Masse, J.-P., Fenerci-Masse, M., 2011. Drowning discontinuities and stratigraphic correlations in platform carbonates. The late Barremian–Early Aptian record of South-East France. Cretaceous Res. 32, 659–684.

Masse, J., Fenerci-Masse, M., 2013. Drowning events, development and demise of carbonate platforms and controlling factors: the Late Barremian–Early Aptian record of Southeast France. Sediment. Geol. 298, 28–52.

Matsukawa, M., Saiki, K., Ito, M., Obata, I., Nichols, D.J., Lockley, M.G., et al., 2006. Early Cretaceous terrestrial ecosystems in East Asia based on food-web and energy-flow models. Cretaceous Res. 27, 285–307.

McGhee Jr., G.R., 1992. Evolutionary biology of the Devonian brachiopoda of New York State no correlation with rate of change of sea level? Lethaia 25, 165–172.

McGhee Jr., G.R., Sheehan, P.M., Bottjer, D.J., Droser, M.L., 2004. Ecological ranking of Phanerozoic biodiversity crises: ecological and taxonomic severities are decoupled. Palaeogeogr Palaeoclimatol. Palaeoecol. 211, 289–297.

McGhee Jr., G.R., Clapham, M.E., Sheehan, P.M., Bottjer, D.J., Droser, M.L., 2013. A new ecological-severity ranking of major Phanerozoic biodiversity crises. Palaeogeogr. Palaeoclimatol Palaeoecol. 370, 260–270.

McRoberts, C.A., Aberhan, M., 1997. Marine diversity and sea-level changes: numerical tests fo association using Early Jurassic bivalves. Int. J. Earth Sci. 86, 160–167.

Melott, A.L., Bambach, R.K., 2011. A ubiquitous $\sim$62-Myr periodic fluctuation superimposed on general trends in fossil biodiversity. II. Evolutionary dynamics associated with periodic fluctuation in marine diversity. Paleobiology 37, 383–408.

Miller, A.I., 1997. Comparative diversification dynamics among palaeocontinents during the Ordovician radiation. Geobios 20, 397–406.

Miller, A.I., Mao, S., 1995. Association of orogenic activity with the Ordovician radiation o marine life. Geology 23, 305–308.

Mitchum Jr., R.M., Van Wagoner, J.C., 1991. High-frequency sequences and their stacking patterns: sequence-stratigraphic evidence of high-frequency eustatic cycles. Sediment. Geol. 70 131–160.

Molina, E., 2015. Evidence and causes of the main extinction events in the Paleogene based o extinction and survival patterns of foraminifera. Earth Sci. Rev. 140, 166–181.

Monnet, C., 2009. The Cenomanian–Turonian boundary mass extinction (Late Cretaceous): new insights from ammonoid biodiversity patterns of Europe, Tunisia and the Western Interior (North America). Palaeogeogr. Palaeoclimatol. Palaeoecol. 282, 88–104.

Nelson, C.S., Hendry, C.H., Jarrett, G.R., 1985. Near-synchroneity of New Zealand Alpine glaciations during the past 750 Kyr. Nature 318, 361–363.

Neraudeau, D., Thierry, J., Moreau, P., 1997. Variation in echinoid biodiversity during the Cenomanian–early Turonian transgressive episode in Charentes (France). Bull. Soc. Geol. Fr. 168, 51–61.

Newell, N.D., 1952. Periodicity of invertebrate evolution. J. Paleontol. 26, 371–385.

Newell, N.D., 1967. Revolutions in the history of life. Geological Society of America Special Paper 89, 63–91.

Newham, E., Benson, R., Upchurch, P., Goswami, A., 2014. Mesozoic mammaliaform diversity: the effect of sampling corrections on reconstructions of evolutionary dynamics. Palaeogeogr. Palaeoclimatol. Palaeoecol. 412, 32–44.

O'Dogherty, L., Sandoval, J., Vera, J.A., 2000. Ammonite faunal turnover tracing sea-level changes during the Jurassic (Betic Cordillera, southern Spain). J. Geol. Soc. 157, 723–736.

Peralta-Medina, E., Falcon-Lang, H.J., 2012. Cretaceous forest composition and productivity inferred from a global fossil wood database. Geology 40, 219–222.

Perrier, V., Meidla, T., Tinn, O., Ainsaar, L., 2012. Biotic response to explosive volcanism: ostracod recovery after Ordovician ash-falls. Palaeogeogr. Palaeoclimatol. Palaeoecol. 365, 166–183.

Perrin, C., Bosellini, F.R., 2012. Paleobiogeography of scleractinian reef corals: changing patterns during the Oligocene–Miocene climatic transition in the Mediterranean. Earth Sci. Rev. 111, 1–24.

Peters, S.E., 2008. Environmental determinants of extinction selectivity in the fossil record. Nature 454, 626–629.

Peters, S.E., Foote, M., 2001. Biodiversity in the Phanerozoic: a reinterpretation. Paleobiology 27, 583–601.

Phelps, R.M., Kerans, C., Da-Gama, R.O.B.P., Jeremiah, J., Hull, D., Loucks, R.G., 2015. Response and recovery of the Comanche carbonate platform surrounding multiple Cretaceous oceanic anoxic events, northern Gulf of Mexico. Cretaceous Res. 54, 117–144.

Poulsen, C.J., Seidov, D., Barron, E.J., Peterson, W.H., 1998. The impact of paleogeographic evolution on the surface oceanic circulation and the marine environment within the mid-Cretaceous Tethys. Paleoceanography 13, 546–559.

Powell, M.G., MacGregor, J., 2011. A geographic test of species selection using planktonic foraminifera during the Cretaceous/Paleogene mass extinction. Paleobiology 37, 426–437.

Powers, C.M., Bottjer, D.J., 2009. The effects of mid-Phanerozoic environmental stress on bryozoan diversity, paleoecology, and paleogeography. Global Planet. Change 65, 146–154.

Prokoph, A., El Bilali, H., Ernst, R., 2013. Periodicities in the emplacement of large igneous provinces through the Phanerozoic: relations to ocean chemistry and marine biodiversity evolution. Geosci. Front. 4, 263–276.

Purdy, E.G., 2008. Comparison of taxonomic diversity, strontium isotope and sea-level patterns. Int. J. Earth Sci. 97, 651–664.

Ramkumar, M., 1996. Occurrence of hardgrounds in the Kallankurichchi Formation (Lower Maestrichtian), Ariyalur Group, Tiruchirapalli Cretaceous sequence, South India and their significance. Indian J. Pet. Geol. 5, 83–97.

Ramkumar, M., 2006. A storm event during the Maastrichtian in the Cauvery Basin, South India. Ann. Geol. Peninsul. Balkan. 67, 35–40.

Ramkumar, M., 2015. Discrimination of tectonic dynamism, quiescence and third order relative sea level cycles of the Cauvery Basin, South India. Ann. Geol. Peninsul. Balkan (in press).

Ramkumar, M., Sathish, G., 2009. Palaeoenvironmental and sequence stratigraphic significance of the occurrence of *Ophiomorpha irregulaire* in the Kallankurichchi Formation, Ariyalur Group, Cauvery Basin, South India. Palaeontol. Stratigr. Facies 17, 129–137.

Ramkumar, M., Stüben, D., Berner, Z., 2004. Lithostratigraphy, depositional history and sea level changes of the Cauvery Basin, South India. Ann. Geol. Peninsul. Balkan. 65, 1–27.

Ramkumar, M., Stüben, D., Berner, Z., 2011. Barremian–Danian chemostratigraphic sequences of the Cauvery Basin, India: implications on scales of stratigraphic correlation. Gondwana Res. 19, 291–309.

Raup, D.M., 1972. Taxonomic diversity during the Phanerozoic. Science, 1065–1071.

Raup, D.M., 1976. Species diversity in the Phanerozoic: a tabulation. Paleobiology 2, 279–288.

Raymo, E., Oppo, D.W., Curry, W., 1997. The mid-Pleistocene climate transition: a deepsea carbon isotopic perspective. Paleoceanography 12, 546–559.

Reolid, M., Sánchez-Quiñónez, C.A., Alegret, L., Molina, E., 2015. Palaeoenvironmental turnover across the Cenomanian–Turonian transition in Oued Bahloul, Tunisia: foraminifera and geochemical proxies. Palaeogeogr. Palaeoclimatol. Palaeoecol. 417, 491–510.

Retallack, G.J., 2011. Exceptional fossil preservation during CO_2 greenhouse crises? Palaeogeogr. Palaeoclimatol. Palaeoecol. 307, 59–74.

Riding, R., 2006. Microbial carbonate abundance compared with fluctuations in metazoan diversity over geological time. Sediment. Geol. 185, 229–238.

Rogers, A.D., 2000. The role of the oceanic oxygen minima in generating biodiversity in the deep sea. Deep Sea Res. Part II 47, 119–148.

Rook, D.L., Heim, N.A., Marcot, J., 2013. Contrasting patterns and connections of rock and biotic diversity in the marine and non-marine fossil records of North America. Palaeogeogr Palaeoclimatol. Palaeoecol. 372, 123–129.

Rose, P.J., Fox, D.L., Marcot, J., Badgley, C., 2011. Flat latitudinal gradient in Paleocene mammal richness suggests decoupling of climate and biodiversity. Geology 39, 163–166.

Rossi, V., Horton, B.P., 2009. The application of a subtidal foraminifera-based transfer function to reconstruct Holocene paleobathymetry of the Po Delta, Northern Adriatic Sea. J Foraminiferal Res. 39, 180–190.

Roy, K., Jablonski, D., Valentine, J.W., Rosenberg, G., 1998. Marine latitudinal diversity gradients: tests of causal hypotheses. Proc. Natl. Acad. Sci. USA 95, 3699–3702.

Ruban, D.A., 2007. Jurassic transgressions and regressions in the Caucasus (northern Neotethy. Ocean) and their influences on the marine biodiversity. Palaeogeogr. Palaeoclimatol. Palaeoecol 251, 422–436.

Ruban, D.A., 2008. Evolutionary rates of the Triassic marine macrofauna and sea-level changes evidences from the Northwestern Caucasus, Northern Neotethys (Russia). Palaeoworld 17 115–125.

Ruban, D.A., 2009. Phanerozoic changes in the high-rank suprageneric diversity structure of brachiopods: linear and non-linear effects. Palaeoworld 18, 263–277.

Ruban, D.A., 2010a. Do new reconstructions clarify the relationships between the Phanerozoi diversity dynamics of marine invertebrates and long-term eustatic trends?. Ann. Paleontol. 96 51–59.

Ruban, D.A., 2010b. Palaeoenvironmental setting (glaciations, sea level, and plate tectonics) o Palaeozoic major biotic radiations in the marine realm. Ann. Paleontol. 96, 143–158.

Sandoval, J., O'Dogherty, L., Guex, J., 2001a. Evolutionary rates of Jurassic ammonites in relation to sea-level fluctuations. Palaios 16, 311–335.

Sandoval, J., O'Dogherty, L., Vera, J.A., Guex, J., 2001b. Sea-level changes and ammonite faunal turnover during the Lias/Dogger transition in the western Tethys. Bull. Soc. Geol. Fr. 173, 57–66.

Schmiedl, G., Kuhnt, T., Ehrmann, W., Emeis, K., Hamann, Y., Kotthoff, U., et al., 2010. Climatic forcing of eastern Mediterranean deep-water formation and benthic ecosystems during the past 22 000 years. Quat. Sci. Rev. 29, 3006–3020.

Sepkoski, J.J., 1981. A factor analytic description of the Phanerozoic marine fossil record. Paleobiology 7, 36–53.

Sepkoski Jr., J.J., 1988. Alpha, beta or gamma: where does all the biodiversity go? Paleobiology 14, 221–234.

Sepkoski Jr., J.J., 2002. A compendium fossil marine animal genera. Bull. Am. Paleontol. 363, 1–560.

Servais, T., Owen, A.W., Harper, D.A.T., Kröger, B., Munnecke, A., 2010. The Great Ordovician Biodiversification Event (GOBE): the palaeoecological dimension. Palaeogeogr. Palaeoclimatol. Palaeoecol. 294, 99–119.

Sessa, J.A., Bralower, T.J., Patzkowsky, M.E., Handley, J.C., Ivany, L.C., 2012. Environmental and biological controls on the diversity and ecology of Late Cretaceous through early Paleogene marine ecosystems in the U.S. Gulf Coastal Plain. Paleobiology 38, 218–239.

Sigwart, J.D., Carey, N., Orr, P.J., 2014. How subtle are the biases that shape the fidelity of the fossil record? A test using marine mollusks. Palaeogeogr. Palaeoclimatol. Palaeoecol. 403, 119–127.

Simberloff, D.S., 1974. Permo-Triassic extinctions: effects of area on biotic equilibrium. J. Geol. 82, 267–274.

Simpson, E.H., 1949. Measurement of diversity. Nature 163, 688.

Smith, A.B., 1994. Systematics and the Fossil Record: Documenting Evolutionary Patterns. Blackwell, Oxford.

Smith, A.B., Gale, A.S., Monks, E.A., 2001. Sea-level change and rock-record bias in the Cretaceous: a problem for extinction and biodiversity studies. Paleobiology 27, 241–253.

Smith, A.B., Monks, N.E.A., Gale, A.S., 2006. Echinoid distribution and sequence stratigraphy in the Cenomanian (Upper Cretaceous) of southern England. Proc. Geol. Assoc. 117, 207–217.

Stanley, S.M., 2007. An analysis of the history of marine animal diversity. Paleobiology 33, 1–55.

Stilwell, J.D., 2003. Patterns of biodiversity and faunal rebound following the K-T boundary extinction event in Austral Palaeocene molluscan faunas. Palaeogeogr. Palaeoclimatol. Palaeoecol. 195, 319–356.

Strother, P.K., 1996. Acritarchs. In: Jansonius, J., McGregor, D.C. (Eds.), Palynology: Principles and Applications, vol. 1. American Association of Stratigraphic Palynologists Foundation, Salt Lake City, UT, pp. 81–106.

Svenning, J.-C., Fløjgaard, C., Marske, K.A., Nógues-Bravo, D., Normand, S., 2011. Applications of species distribution modeling to paleobiology. Quat. Sci. Rev. 30, 2930–2947.

Tappan, H., 1968. Primary production, isotopes, extinctions and the atmosphere. Palaeogeogr. Palaeoclimatol. Palaeoecol. 4, 187–210.

Tappan, E., Loeblich Jr., A.R., 1973. Evolution of the oceanic plankton. Earth Sci. Rev. 9, 207–240.

Tibert, N.E., Leckie, R.M., 2004. High-resolution estuarine sea level cycles from the late Cretaceous: amplitude constraints using agglutinated foraminifera. J. Foraminiferal Res. 34, 130–143.

Toshimitsu, S., Hirano, H., Matsumoto, T., Takahashi, K., 2003. Database and species diversity of Japanese Cretaceous ammonoids. J. Asian Earth Sci. 21, 887–893.

Tucker, M.E., Gallagher, J., Leng, M.J., 2009. Are beds in shelf carbonates millennial-scale cycles? An example from the mid-Carboniferous of northern England. Sediment. Geol. 214, 19–34.

Ullmann, C.V., Frei, R., Korte, C., Hesselbo, S.P., 2015. Chemical and isotopic architecture of the belemnite rostrum. Geochim. Cosmochim. Acta 159, 231–243.

Vail, P.R., Mitchum, R.M., Thompson, S., 1977. Seismic stratigraphy and global changes of sea level. Part 4. Global cycles of relative changes of sea level. In: Payton, C.E. (Ed.), Seismic Stratigraphy—Applications to Hydrocarbon Exploration, vol. 26. Memoirs of American Association of Petroleum Geologists, pp. 83–97.

Valentine, J.W., 1968. Climatic regulation of species diversification and extinction. Bull. Geol. Soc. Am. 79, 273–276.

Vandenbroucke, T.R.A., Emsbo, P., Munnecke, A., Nuns, N., Duponchel, L., Lepot, K., et al., 2015. Metal-induced malformations in early Palaeozoic plankton are harbingers of mass extinction. Nat. Commun. Available from: http://dx.doi.org/10.1038/ncomms8966.

Vecoli, M., Hérissé, A.L., 2004. Biostratigraphy, taxonomic diversity and patterns of morphological evolution of Ordovician acritarchs (organic-walled microphytoplankton) from the northern Gondwana margin in relation to palaeoclimatic and palaeogeographic changes. Earth Sci. Rev. 67, 267–311.

Veizer, J., 1985. Carbonates and ancient oceans: isotopic and chemical record on timescales of 10^7–10^9 years. In: Sunquist, E.T., Broecker, W.S. (Eds.), The Carbon Cycle and Atmospheric CO_2: Natural Variations—Archaen to Present, vol. 32. American Geophysical Union Monograph, pp. 595–601.

Veizer, J., Bruckschen, P., Pawellek, F., 1997. Oxygen isotope evolution of Phanerozoic seawater. Palaeogeogr. Palaeoclimatol. Palaeoecol. 132, 159–172.

Veizer, J., Ala, D., Azmy, K., 1999. $^{87}Sr/^{86}Sr$, $\delta^{13}C$ and $\delta^{18}O$ evolution of Phanerozoic seawater. Chem. Geol. 161, 59–88.

Veizer, J., Godderis, Y., Francois, L.M., 2000. Evidence for decoupling of atmospheric CO_2 and global climate during the Phanerozoic eon. Nature 408, 698–701.

Villier, L., Navarro, N., 2004. Biodiversity dynamics and their driving factors during the Cretaceous diversification of Spatangoida (Echinoidea, Echinodermata). Palaeogeogr. Palaeoclimatol. Palaeoecol. 214, 265–282.

Wang, Y.-D., Axsmith, B.J., 2008. Biodiversity, anatomy and evolution of Mesozoic plants: an introduction. Palaeoworld 17, 163–165.

Williams, D.F., Thunell, R.C., Tappa, E., 1988. Chronology of the Pleistocene oxygen isotope record 0–1.88 MY BP. Palaeogeogr. Palaeoclimatol. Palaeoecol. 64, 221–240.

Yacobucci, M.M., 2005. Multifractal and white noise evolutionary dynamics in Jurassic–Cretaceous Ammonoidea. Geology 33, 97–100.

Printed in the United States
By Bookmasters